Advanced Techniques in Geospatial Data Analysis and Mapping with Python

by Edward Franklin

ISBN: 978-1-77890-389-2
Imprint: Montecito Hot Springs

Contents

Introduction

Background of Geospatial Data Analysis

Geospatial data analysis is a field that focuses on extracting meaningful information from spatial datasets. It involves the use of computational techniques to analyze and interpret geospatial data for various applications such as urban planning, environmental management, public health, and disaster management. In this section, we will provide an overview of the background and key principles of geospatial data analysis.

Principles of Geospatial Data Analysis

Geospatial data analysis encompasses a range of techniques and methods from multiple disciplines, including geography, spatial statistics, remote sensing, and geographic information systems (GIS). These disciplines provide the foundation for understanding and analyzing spatial patterns and relationships in data.

One fundamental principle of geospatial data analysis is the concept of spatial autocorrelation. Spatial autocorrelation refers to the tendency of similar values to occur in close proximity to each other in space. This principle is based on the assumption that nearby locations are more likely to have similar attributes than distant locations.

Spatial autocorrelation can be measured using various statistical methods, such as Moran's I. Moran's I is a measure of spatial autocorrelation that quantifies the degree of similarity between spatial units based on their attribute values. It ranges from -1 (indicating perfect dispersion) to 1 (indicating perfect clustering). Positive values of Moran's I indicate positive spatial autocorrelation, meaning that similar values tend to cluster together, while negative values indicate negative spatial autocorrelation, meaning that dissimilar values tend to be adjacent to each other.

Another principle of geospatial data analysis is spatial clustering. Spatial clustering refers to the identification and characterization of groups or clusters of similar spatial units. This technique allows us to identify areas with similar attributes or behaviors, which can be useful for various applications such as identifying high-crime areas or disease clusters.

There are several spatial clustering techniques available, including K-means clustering and hierarchical clustering. K-means clustering is an unsupervised machine learning algorithm that partitions a dataset into a predetermined number of clusters. It seeks to minimize the within-cluster sum of squared Euclidean distances. Hierarchical clustering, on the other hand, creates a hierarchy of clusters by iteratively merging or splitting clusters based on a specified distance metric.

Geospatial data analysis also involves the use of geostatistics, which is a set of statistical techniques used to analyze spatially correlated data. Geostatistics takes into account the spatial structure of the data and provides a means for making predictions or estimates at unobserved locations. Ordinary Kriging is one commonly used geostatistical method that predicts values at unsampled locations based on the spatial autocorrelation of the data.

Challenges in Geospatial Data Analysis

Geospatial data analysis presents several challenges that need to be addressed to ensure accurate and meaningful results. One challenge is data quality assurance, as spatial datasets are prone to errors and inconsistencies. Issues such as missing data, outliers, and measurement errors can significantly affect the analysis outcomes. Therefore, data cleaning and preprocessing techniques are essential to identify and correct these issues.

Another challenge is data integration and interoperability. Geospatial datasets often come from multiple sources and formats, making it necessary to integrate and harmonize them for analysis. Data integration involves combining multiple datasets into a single consistent format, while data interoperability ensures that different software tools and systems can exchange and interpret spatial data.

The handling of big geospatial data is another challenge in geospatial data analysis. With the advent of remote sensing technologies and the increasing availability of high-resolution imagery, geospatial datasets have grown exponentially in size. Analyzing large-scale spatial datasets requires efficient computational techniques and algorithms capable of processing and analyzing the data within reasonable time constraints.

Ethical Considerations in Geospatial Data Analysis

Geospatial data analysis raises ethical considerations related to data privacy, confidentiality, and bias. Spatial datasets often contain sensitive information about individuals, such as their location or socioeconomic status. It is crucial to handle this information responsibly and ensure that data anonymization techniques are applied to protect privacy.

Another ethical consideration is the potential for biases in geospatial analysis. Biases can arise from various sources, including sampling biases in data collection or algorithmic biases in data processing. Biases can lead to erroneous conclusions or perpetuate existing social inequalities. It is important to be aware of these biases and employ techniques to mitigate or correct them.

Resources and Further Reading

To delve deeper into geospatial data analysis, the following resources are recommended:

- *Spatial Analysis: A Guide for Ecologists* by Fortin and Dale is an excellent introductory book that covers the principles of geospatial data analysis and spatial statistics.

- *Geospatial Analysis: A Comprehensive Guide* by de Smith, Goodchild, and Longley provides a comprehensive overview of geospatial analysis techniques and their applications.

- *Python Geospatial Analysis Cookbook* by Umbehr and Lawhead offers practical examples and code snippets for performing geospatial analysis using Python.

- *The ESRI Guide to GIS Analysis, Volume 2: Spatial Measurements and Statistics* by Mitchell explores various spatial statistics techniques for analyzing geospatial data using ESRI software.

In addition to these resources, there are several online tutorials, forums, and community websites dedicated to geospatial data analysis, such as GeoNet and the GIS Stack Exchange, where you can find further guidance and support.

Conclusion

In this section, we have provided an overview of the background and key principles of geospatial data analysis. We discussed the principles of spatial autocorrelation and spatial clustering, as well as the challenges and ethical considerations in geospatial analysis. Understanding these principles and addressing the associated challenges is essential for conducting accurate and meaningful geospatial data analysis. In the next section, we will explore the various data acquisition and preprocessing techniques necessary for geospatial analysis.

Data Acquisition and Preprocessing

Data Sources

In the field of geospatial data analysis, obtaining reliable and accurate data is crucial for conducting meaningful analyses and generating insightful results. This section focuses on the various sources from which geospatial data can be acquired. We will explore both traditional and emerging sources, highlighting their strengths, limitations, and considerations.

Traditional Sources

Traditionally, geospatial data sources have primarily consisted of government agencies, research institutions, and commercial data providers. These sources offer well-curated and authoritative datasets that are often used as the basis for various analyses.

Government Agencies

Government agencies play a vital role in providing geospatial data that is collected and maintained for administrative purposes. These agencies include national mapping agencies, environmental departments, statistical bureaus, and transportation authorities.

For example, the United States Geological Survey (USGS) offers a wealth of geospatial data, including digital elevation models, land use maps, and hydrological data. Similarly, the National Aeronautics and Space Administration (NASA) provides satellite imagery and climate data.

While government data sources are generally reliable and comprehensive, they may have limitations. These limitations can include data availability, resolution, and timeliness. Additionally, data from different government agencies may have inconsistencies and variations in format and standards, requiring careful preprocessing before analysis.

Research Institutions

Research institutions, such as universities and scientific organizations, often generate geospatial datasets through extensive field surveys and experiments. These datasets are particularly valuable for specialized research and projects that require hyper-local or detailed information. For example, a university conducting an ecological study might collect data on vegetation cover, soil composition, and animal habitats.

Research institution data sources may provide valuable insights for niche projects, but they are typically more limited in coverage and scale compared to government datasets. The collection process can also be time-consuming and expensive, resulting in datasets that are not as regularly updated or widely available.

Commercial Data Providers

Commercial data providers have emerged as important sources of geospatial data, offering a wide range of datasets for various industries and applications. These providers collect, process, and distribute data from diverse sources, including aerial and satellite imagery, demographic data, and business location data.

Commercial datasets are often of high quality, with detailed attributes and extensive coverage. They can be readily integrated into analyses and applications. However, commercial data is not always freely available, and licensing or subscription fees may apply. Additionally, the metadata and data update frequency should be carefully considered, as commercial datasets may lag behind the most recent changes on the ground.

Emerging Sources

In recent years, with advancements in technology and the ever-increasing availability of data, new sources of geospatial data have emerged. These emerging sources offer exciting opportunities but also come with their own set of challenges.

Crowdsourced Data

Crowdsourced data refers to information contributed by the general public or a specific community. This type of data collection leverages the power of citizen science and user-generated content. Platforms such as OpenStreetMap (OSM) allow users to contribute and edit mapping data, creating a collaborative and constantly updated geospatial dataset.

Crowdsourced data provides a valuable source of information, especially in areas where authoritative datasets are lacking or outdated. It enables community participation and can capture rapidly changing or localized phenomena. However, the quality and reliability of crowdsourced data can vary, requiring careful validation and integration. Biases and inaccuracies may arise due to differences in data collection methodologies and individual contributors' expertise.

Sensor Networks and Internet of Things (IoT)

Advancements in sensor technologies and the rise of the Internet of Things (IoT) have led to the proliferation of real-time geospatial data from various sources. These include weather stations, air quality sensors, traffic monitoring devices, and other IoT-enabled devices. The data collected from these sources can provide valuable insights into dynamic phenomena, such as weather patterns, air pollution levels, and transportation patterns.

Sensor-generated data offers near-real-time information, enabling timely decision-making and monitoring of dynamic systems. However, challenges exist in terms of data quality assurance, compatibility, and scalability of sensor networks. Additionally, issues related to data privacy and security need to be addressed when dealing with sensitive data collected from IoT devices.

Considerations and Best Practices

When selecting and utilizing geospatial data sources for analysis, several factors should be considered:

Data Quality

Assessing the quality of geospatial data is essential to ensure the accuracy and reliability of the analysis results. Factors such as positional accuracy, attribute completeness, and temporal consistency should be evaluated. Understanding the data's lineage and metadata documentation is crucial for making informed decisions about its suitability for specific applications.

Data Format and Standards

Geospatial data comes in various formats, including shapefiles, raster images, and geoJSON. Understanding the format and the associated standards is essential for effective data integration and analysis. It is also important to consider data interoperability and conversion requirements when working with datasets from different sources.

Data Licensing and Intellectual Property

Before using geospatial data, it is important to understand and adhere to any licensing restrictions or intellectual property rights. Some datasets may be openly available, while others require permission or appropriate licensing agreements. Compliance with legal requirements ensures ethical and responsible use of the data.

Data Ethics and Privacy

In the era of big data, responsible data use and protecting individual privacy are paramount. Geospatial data often contains sensitive information about individuals or identifiable locations. Adhering to privacy regulations, anonymizing data, and implementing robust data security measures are essential when working with geospatial data.

Case Study: Urban Planning and Data Sources

To illustrate the importance of data sources in a real-world context, let's consider an example in urban planning. In the process of developing a new city park, urban planners rely on various geospatial data sources.

Government data sources provide foundational data such as land use, zoning regulations, and transportation networks. This information allows planners to identify suitable locations for the park and consider environmental factors. Research institution data may provide valuable insights into local biodiversity, habitat preservation, and the impact on the surrounding ecosystem.

Commercial data providers can contribute demographic data, providing information about the local population and their recreational preferences. Moreover, crowdsourced data, such as feedback from residents or public consultations, enables community engagement and helps shape the park design to meet the needs of the people.

Sensor networks and IoT devices can play a role in monitoring air quality, noise levels, and usage patterns within existing parks, allowing planners to optimize the park's amenities for the well-being of park-goers.

By integrating and analyzing data from multiple sources, urban planners can make informed decisions, ensuring the new park's design aligns with the community's needs and enhances the overall urban environment.

Summary

This section explored the various sources of geospatial data, including traditional and emerging sources. Traditional sources, such as government agencies, research institutions, and commercial data providers, offer authoritative and comprehensive datasets. Emerging sources, including crowdsourced data and sensor networks, provide dynamic and real-time information. Considerations such as data quality, format, licensing, ethics, and privacy should guide the selection and utilization of geospatial data sources. By utilizing a combination of data sources, geospatial analysts can gain valuable insights for conducting meaningful analyses and solving real-world problems.

Resources and Further Reading

To delve deeper into the topic of geospatial data sources, the following resources are recommended:

- Longley, P. A., Goodchild, M. F., Maguire, D. J., & Rhind, D. W. (2018). *Geographic Information Science & Systems*. Wiley.
- Arctur, D., & Zeiler, M. (2020). *Spatial Data Infrastructures in Context: North and South*. Esri Press.
- Goodchild, M. F., & Janelle, D. G. (Eds.). (2004). *Spatially Integrated Social Science*. Oxford University Press.
- OpenStreetMap: `https://www.openstreetmap.org/`

Remember to explore the websites of government agencies, research institutions, and commercial data providers specific to your area of interest, as they often provide access to relevant geospatial datasets.

Data Cleaning and Quality Assurance

Introduction

In the field of geospatial data analysis, it is crucial to ensure the quality and reliability of the data used for analysis. Raw data obtained from various sources can be messy, contain errors, or have inconsistencies. Therefore, data cleaning and quality assurance techniques play a vital role in preparing the data for further analysis. This section focuses on the importance of data cleaning and quality assurance in geospatial data analysis and provides practical techniques to address common data quality issues.

Data Quality Issues

Data quality refers to the accuracy, completeness, consistency, and timeliness of the data. In geospatial data analysis, several common data quality issues need to be addressed:

- **Missing Data:** Missing data points are gaps in the dataset, which can occur due to various reasons such as sensor failures, data transmission errors, or incomplete field surveys. Missing data can affect the accuracy of analysis results and need to be handled appropriately.
- **Outliers:** Outliers are extreme values that deviate significantly from the expected pattern. Outliers can occur due to measurement errors, data entry mistakes, or rare events. Outliers can distort statistical analysis and modeling results and should be identified and treated properly.
- **Inconsistent Data:** Inconsistencies in geospatial data can arise when different datasets or data sources contain conflicting information. Inconsistent data can lead to inaccurate analysis and generate misleading results. It is essential to detect and resolve inconsistencies to ensure data integrity.
- **Data Accuracy:** Data accuracy refers to the agreement between the measured or observed values and the true values. Geospatial datasets often have inherent errors introduced during data collection and processing. These errors need to be quantified and corrected to improve the reliability of analysis results.
- **Spatial Bias:** Spatial bias refers to the presence of systematic errors or biases related to the spatial nature of the data. Spatial bias can occur due to uneven distribution

of data collection locations, differences in measurement techniques, or environmental factors. Correcting and addressing spatial biases is essential for unbiased geospatial analysis.

Data Cleaning Techniques

Data cleaning involves identifying and rectifying the data quality issues discussed earlier. Various techniques can be used to clean geospatial data:

- **Missing Data Imputation:** When dealing with missing data, imputation techniques can be used to estimate the missing values based on available information. Simple techniques such as mean imputation or linear regression imputation can be employed. However, more advanced methods like k-nearest neighbors imputation or multiple imputation can be used for more accurate results.

- **Outlier Detection and Treatment:** Outliers need to be identified and either removed or adjusted to avoid their influence on subsequent analysis. Statistical techniques such as the Z-score or the interquartile range (IQR) can be used to detect outliers. Depending on the nature of the data, outliers can be replaced with more reasonable values or entirely removed.

- **Data Integration and Cleaning:** In cases where multiple datasets are used, data integration techniques are required to merge them into a consistent format. This process involves resolving naming inconsistencies, standardizing units of measurement, and aggregating or disaggregating data as needed.

- **Addressing Data Accuracy:** Data accuracy issues can be addressed through calibration or validation techniques.

Calibration involves correcting systematic errors in the data by applying correction factors or interpolation techniques. Validation, on the other hand, compares the data against independent reference data to evaluate its accuracy.

Quality Assurance

Quality assurance (QA) is a comprehensive process to ensure that the data meets specific standards and requirements. It involves a set of activities performed throughout the lifecycle of data analysis. The primary objectives of QA in geospatial data analysis are as follows:

- **Data Validation:** Data validation is the process of assessing data to ensure that it is accurate, complete, consistent, and reliable. This involves performing various checks, such as range checks, cross-validation, and logical checks, to identify and correct any potential errors or inconsistencies.

- **Metadata Management:** Metadata provides essential information about the data, including its source, format, and description. Managing metadata ensures that data remains well-documented, making it easier to understand, track changes, and share with other analysts.

- **Version Control:** Version control is crucial for tracking changes made to the data and preventing unintentional modifications or loss. By maintaining a versioning system, it becomes easier to trace back to a specific version of the data if any issues arise during the analysis process.

- **Documentation and Annotation:** Documenting the data processing steps, assumptions, and transformations is essential for reproducibility and transparency. Annotation provides contextual information, such as definitions of variables and data sources, enabling others to understand and interpret the data correctly.

Best Practices and Challenges

To ensure effective data cleaning and quality assurance in geospatial data analysis, the following best practices should be considered:

- **Standardization:** Standardize data formats, units of measurement, and naming conventions to ensure consistency and facilitate data integration.
- **Automated Data Cleaning:** Utilize automated scripts or workflows to perform routine data cleaning tasks, reducing errors and improving efficiency.
- **Continuous Monitoring:** Implement mechanisms for continuous monitoring of data quality to detect issues early and prevent their propagation throughout the analysis.
- **Validation with Ground Truth:** Whenever possible, validate the accuracy of the geospatial data by comparing it with ground truth data collected through independent means.
- **Collaborative Data Cleaning:** Encourage collaboration between data providers, analysts, and domain experts to gain a comprehensive understanding of the data and address any quality issues collectively.
- **Data Quality Reporting:** Develop clear and concise reports that summarize the findings and outcomes of data cleaning and quality assurance processes. These reports can aid in decision-making and provide transparency to stakeholders.

Conclusion

Data cleaning and quality assurance are indispensable steps in geospatial data analysis. By addressing data quality issues and ensuring the reliability of the data, analysts can produce accurate and trustworthy analysis results. This section provided an overview of data cleaning techniques, quality assurance practices, and challenges in geospatial data analysis. Incorporating these techniques and best practices will enhance the integrity and reliability of geospatial analysis results and enable better decision-making in various domains.

Data Integration and Interoperability

In the field of geospatial data analysis, data integration and interoperability play a crucial role in ensuring that geospatial datasets from different sources and formats can be effectively combined and analyzed. With the increasing availability of diverse data sources and the need for comprehensive analyses, it is essential to have mechanisms in place for integrating and harmonizing these datasets.

Background

Geospatial data comes in various formats, including raster and vector data, and is often collected from different sources such as satellite imagery, GPS devices, and remote sensing techniques. Additionally, geospatial data can be stored in various data formats, such as shapefiles, GeoJSON, raster data formats (e.g., TIFF, GeoTIFF), and spatial databases (e.g., PostGIS). Therefore, there is a need for techniques that allow for the integration and interoperability of these diverse datasets.

Data Integration Techniques

Data integration involves combining data from multiple sources to create a unified view of the data. In geospatial data analysis, several techniques are used for data integration:

Spatial Join

Spatial join is a common technique used to integrate geospatial datasets by combining their attribute information based on the spatial relationship between their geometries. For example, suppose we have a dataset of cities and another dataset of countries. By performing a spatial join operation, we can associate each city with the corresponding country based on their spatial intersection.

Data Fusion

Data fusion is the process of combining multiple datasets to create a new dataset that contains information from all the input datasets. This technique is particularly useful when working with multiple data sources that provide complementary information. For example, suppose we have satellite imagery data and ground truth survey data. By fusing these datasets, we can create a more accurate and detailed dataset that captures both the high-level overview from satellite imagery and the precise details from ground truth surveys.

Data Wrangling

Data wrangling involves the transformation and preparation of data to make it suitable for analysis. This technique is essential when working with datasets in different formats or with inconsistencies in their structure or content. Data wrangling techniques, such as data cleaning, data transformation, and data enrichment, are used to harmonize datasets before integrating them.

Data Interoperability

Data interoperability refers to the ability of different systems and software to exchange and interpret data seamlessly. In geospatial data analysis, ensuring data interoperability is crucial for sharing and integrating datasets across different platforms and software tools.

Standardization

Standardization is a fundamental principle for achieving data interoperability. By establishing common data standards, formats, and schemas, it becomes easier for different systems and software tools to understand and process geospatial data. Standards such as GeoJSON, OGC standards (e.g., WMS, WFS), and geospatial databases (e.g., PostGIS) enable interoperability by providing a common framework for data exchange and analysis.

Open Geospatial Consortium (OGC) Standards

The Open Geospatial Consortium (OGC) is an international consortium that develops and promotes standards for geospatial data interoperability. OGC standards define common protocols and formats for data exchange, including standards for web mapping (WMS), web feature services (WFS), and web coverage services (WCS). Adhering to OGC standards ensures that geospatial data can be accessed and processed by different systems and software tools.

Challenges and Solutions

Integrating and interoperating geospatial data can present several challenges. Some common challenges include differences in data formats, varying levels of data quality, and the need to handle large volumes of data. These challenges can be addressed through the following solutions:

Data Transformation

Data transformation techniques, such as data conversion and reformatting, can be used to convert data from one format to another. For example, if we have data in shapefile format and need to integrate it with data in GeoJSON format, we can use data transformation techniques to convert the data into a compatible format.

Data Quality Assurance

Data quality assurance techniques, such as data cleaning and validation, are used to identify and correct data inconsistencies, errors, and outliers. By ensuring data quality, the integrated dataset becomes more reliable and accurate for analysis.

Spatial Indexing

Spatial indexing techniques, such as quadtree and R-tree indexing, can improve the performance of spatial queries by organizing geospatial data in a hierarchical structure. Spatial indexing enables efficient retrieval and analysis of specific subsets of the integrated dataset, reducing query response times for large volumes of data.

Case Study: Integration of Satellite Imagery and Climate Data

Let's consider a case study where we want to integrate satellite imagery data with climate data to analyze the relationship between land surface temperature and vegetation cover. The satellite imagery data provides information on land surface temperature at a given spatial resolution, while the climate data provides information on vegetation cover.

To integrate these datasets, we can use a combination of spatial join and data fusion techniques. First, we perform a spatial join operation to associate the climate data with the corresponding spatial locations in the satellite imagery dataset. This allows us to link the temperature information with the vegetation cover information.

Next, we apply data fusion techniques to combine the temperature and vegetation cover data into a new dataset. This dataset will contain both temperature and vegetation cover attributes for each spatial location, providing a comprehensive view of the relationship between these variables.

By analyzing this integrated dataset, we can uncover insights such as the impact of vegetation cover on land surface temperature and identify areas where vegetation cover plays a significant role in temperature regulation.

Conclusion

In geospatial data analysis, data integration and interoperability are essential for combining and analyzing diverse datasets effectively. Techniques such as spatial join, data fusion, and data wrangling enable the integration of datasets from different sources and formats. Data interoperability, achieved through standardization and adherence to OGC standards, allows for seamless data exchange and analysis across different systems and software tools. However, challenges such as data format differences and data quality variations need to be addressed through data transformation, quality assurance, and spatial indexing techniques. Through careful integration and interoperability, geospatial analysts can gain valuable insights from combining disparate datasets and contribute to a more comprehensive understanding of the world around us.

Resources and Further Reading

1. Goodchild, M. F., & Janelle, D. G. (Eds.). (2004). Spatially Integrated Social Science. Oxford University Press. 2. Longley, P. A., Goodchild, M. F., Maguire, D. J., & Rhind, D. W. (2015). Geographic Information Science & Systems. John Wiley & Sons. 3. Open Geospatial Consortium (OGC) Standards:

https://www.ogc.org/standards 4. PostGIS: https://postgis.net/ 5. Python GIS Tools: https://github.com/geopandas/geopandas 6. QGIS: https://qgis.org/ 7. Woo, W., Adams, B. J., & Dwivedi, R. (2010). Environmental modelling with GIS and remote sensing. CRC Press.

Exercises

1. Identify two data integration techniques and explain their applications in geospatial data analysis. 2. What are the benefits of adhering to OGC standards for geospatial data interoperability? 3. How can data quality assurance techniques improve the reliability of integrated datasets? 4. Explain the concept of spatial indexing and its role in improving query performance. 5. Choose a real-world application (e.g., transportation planning, biodiversity conservation) and discuss how data integration and interoperability can contribute to the analysis and decision-making process. 6. Explore the PostGIS documentation and find an example of a spatial join operation using SQL. 7. Discuss the potential ethical considerations and challenges related to the integration and interoperability of geospatial data.

Remember to refer to the resources provided and conduct further research to enhance your understanding of data integration and interoperability in geospatial data analysis.

Exploratory Data Analysis

Descriptive Statistics and Data Visualization

In this section, we will explore the fundamental concepts and techniques of descriptive statistics and data visualization in geospatial data analysis. Descriptive statistics provide us with a summary of the main characteristics of a dataset, while data visualization allows us to visually represent and interpret the data. These techniques are essential for gaining insights into the spatial patterns and distributions present in geospatial datasets.

Introduction to Descriptive Statistics

Descriptive statistics is the branch of statistics that focuses on organizing, summarizing, and presenting data in a meaningful way. It provides us with measures that describe the central tendency, variability, and shape of a dataset. These measures can be applied to both numerical and categorical data.

Some of the key concepts in descriptive statistics include:

- **Measures of central tendency:** These measures indicate

the center or average of a dataset. The most commonly used measures of central tendency are the mean, median, and mode.

- **Measures of variability:** These measures reflect the spread or dispersion of a dataset. They include the range, variance, and standard deviation.

- **Measures of shape:** These measures provide information about the shape of the distribution of a dataset. Skewness measures the asymmetry of the distribution, while kurtosis measures the heaviness of the tails.

- **Percentiles and quartiles:** Percentiles divide a dataset into equal portions, while quartiles divide it into four equal parts. These measures help understand the distribution of the dataset and identify outliers.

Data Visualization Techniques

Data visualization is the graphical representation of data to uncover patterns, relationships, and trends. It helps in better understanding and interpretation of the data by presenting information visually. In geospatial data analysis, data visualization is particularly crucial as it allows us to explore the spatial distribution and patterns of phenomena.

There are various types of data visualization techniques that can be used depending on the nature of the data and the analysis objectives. Some common techniques include:

- **Bar charts:** Bar charts are used to present categorical data. They consist of vertical or horizontal bars that represent different categories and their corresponding frequencies or proportions.

- **Histograms:** Histograms are used to display the distribution of numerical data. They consist of adjacent rectangles or bins, where the area of each bin represents the frequency or proportion of data falling within that range.
- **Scatter plots:** Scatter plots are used to visualize the relationship between two numerical variables. Each data point is represented as a dot on the graph, with the x-axis and y-axis representing the two variables.
- **Choropleth maps:** Choropleth maps are used to display spatially aggregated data. Different regions or areas are colored or patterned according to the values or categories they represent.
- **Heatmaps:** Heatmaps are used to represent the intensity or density of a phenomenon across space. They use a color gradient to show variations in values across a geographic area.

Descriptive Statistics and Data Visualization in Python

Python offers a wide range of libraries and tools for descriptive statistics and data visualization in geospatial analysis. Some popular libraries include NumPy, Pandas, Matplotlib, and Seaborn. These libraries provide a comprehensive set of functions and methods for calculating descriptive statistics and creating visually appealing plots.

To calculate various measures of central tendency, variability, and shape, you can use functions provided by the NumPy and Pandas libraries. These libraries also offer methods to compute percentiles and quartiles.

For data visualization, Matplotlib is a powerful library that provides a wide range of plot types, including bar charts, histograms, scatter plots, choropleth maps, and heatmaps. Matplotlib allows for extensive customization, enabling the creation of publication-quality visualizations.

Seaborn is another popular library that builds on top of Matplotlib and provides a higher-level interface for creating aesthetically pleasing statistical visualizations. It simplifies the process of creating complex plots and offers additional functionalities, such as automatic data binning and handling of missing data.

Example: Descriptive Statistics and Data Visualization

Let's consider a geospatial dataset containing information about the population density of different cities in a country. We want to analyze the dataset using descriptive statistics and visualize the population density distribution across the country.

First, we can calculate the mean, median, and standard deviation of the population density using the NumPy library:

$$\text{mean} = \text{np.mean(population\_density)}$$

$$\text{median} = \text{np.median(population\_density)}$$

$$\text{std\_dev} = \text{np.std(population\_density)}$$

Next, we can create a histogram to visualize the distribution of population density:

plt.hist(population_density, bins=10, edgecolor='black')

To create a choropleth map that shows the population density across the country, we can utilize the geospatial capabilities of libraries like Geopandas and GeoPandas:

```
geopandas_data.plot(column='population_density', cmap='YlOrRd', line
```

This example showcases how descriptive statistics and data visualization techniques can be applied to gain insights into geospatial datasets.

Resources and Further Reading

To further explore the concepts and techniques discussed in this section, refer to the following resources:

- *Python for Data Analysis* by Wes McKinney
- *Data Visualization with Python and Matplotlib* by Valentina Alto
- *Python Data Science Handbook* by Jake VanderPlas
- *Geospatial Data Analysis and Visualization with Python* by Gerard Toonstra
- Online tutorials and documentation for NumPy, Pandas, Matplotlib, Seaborn, Geopandas, and GeoPandas

These resources provide in-depth explanations, examples, and practical guidance on descriptive statistics and data visualization in geospatial data analysis.

Spatial Autocorrelation Analysis

Spatial autocorrelation refers to the degree of similarity or dissimilarity of a variable in relation to its neighboring locations in space. It measures the spatial patterns of a variable to determine if there is clustering or dispersion of values. In this section, we will explore the concept of spatial autocorrelation and discuss techniques to measure and interpret it.

Introduction to Spatial Autocorrelation

Spatial autocorrelation is based on the principle that nearby locations tend to have similar values, while distant locations tend to have different values. This phenomenon is often observed in geospatial datasets, where spatially clustered patterns are common. Understanding spatial autocorrelation is crucial for various fields, including urban planning, epidemiology, ecology, and environmental studies.

Spatial autocorrelation can be described using the Moran's I statistic. Moran's I is a measure of spatial autocorrelation that quantifies the similarity or dissimilarity between the values of a variable at different locations. It ranges from -1 to 1, where values close to 1 indicate strong positive spatial autocorrelation (similar values are close to each other), values close to -1 indicate strong negative spatial autocorrelation (dissimilar values are close to each other), and values close to 0 indicate no spatial autocorrelation.

Techniques for Measuring Spatial Autocorrelation

There are several techniques available to measure spatial autocorrelation. We will discuss two commonly used methods: Moran's I and Geary's C.

Moran's I

Moran's I measures spatial autocorrelation by comparing each location to its neighboring locations. The formula for Moran's I is as follows:

$$I = \frac{n}{W} \cdot \frac{\sum_i \sum_j w_{ij}(x_i - \bar{x})(x_j - \bar{x})}{\sum_i (x_i - \bar{x})^2}$$

where:

- n is the number of locations

- W is the sum of the weights (w_{ij}) between all pairs of locations
- x_i and x_j are the values at locations i and j, respectively
- $\bar{x}$ is the mean value of the variable across all locations

To interpret Moran's I, we compare it to the expected value under spatial randomness. If the calculated Moran's I is significantly different from the expected value, there is spatial autocorrelation present. The significance can be tested using permutation tests or by calculating a z-score.

Geary's C

Geary's C is another measure of spatial autocorrelation that focuses on the differences between neighboring locations. The formula for Geary's C is as follows:

$$C = \frac{(n-1)}{2 \cdot W} \cdot \frac{\sum_i \sum_j w_{ij}(x_i - x_j)^2}{\sum_i (x_i - \bar{x})^2}$$

where variables have the same meaning as in Moran's I.

Geary's C ranges from 0 to 2, with values closer to 0 indicating strong positive spatial autocorrelation, values closer to 2 indicating strong negative spatial autocorrelation, and values around 1 indicating no spatial autocorrelation.

Interpretation of Spatial Autocorrelation

Once we have calculated Moran's I or Geary's C, we need to interpret the results to understand the spatial patterns. The interpretation depends on the value of the statistics and the associated p-value.

If the statistic is significantly different from the expected value at a certain confidence level (e.g., $p < 0.05$), we can conclude that there is spatial autocorrelation. A positive statistic indicates positive spatial autocorrelation, meaning that similar values tend to cluster together. A negative statistic indicates negative spatial autocorrelation, meaning that dissimilar values tend to cluster together.

If the statistic is not significantly different from the expected value, we can conclude that there is no spatial autocorrelation. However, it is important to note that the absence of significant autocorrelation does not mean the absence of spatial patterns. Other techniques, such as exploratory spatial data analysis and spatial clustering, should be used to further investigate the dataset.

Applications and Examples

Spatial autocorrelation analysis has various applications in different fields. Let's explore a few examples:

Urban Planning

In an urban planning study, spatial autocorrelation can help identify areas with similar characteristics, such as income levels or crime rates. This information can guide the allocation of resources and target interventions more effectively.

Epidemiology

Spatial autocorrelation is widely used in epidemiology to understand the spread of diseases. By analyzing the spatial patterns of disease occurrences, researchers can identify clusters and determine if there are any underlying socio-environmental factors contributing to disease transmission.

Ecology

In ecology, spatial autocorrelation analysis allows researchers to study the distribution of species and the factors influencing their distribution. It helps identify hotspots of biodiversity, understand habitat connectivity, and design effective conservation strategies.

Practical Considerations and Caveats

When performing spatial autocorrelation analysis, there are several important considerations to keep in mind:

- Selection of appropriate spatial scale: The analysis should be conducted at an appropriate spatial scale that is relevant to the research question and the phenomenon being studied. Different scales may reveal different patterns of spatial autocorrelation.

- Handling spatial heterogeneity: Spatial autocorrelation analysis assumes stationarity, meaning that the relationship between neighboring locations is consistent across the study area. However, spatial heterogeneity can violate this assumption. It is important to account for spatial heterogeneity to obtain accurate results.

- Data preprocessing: Data should be preprocessed to ensure data quality, handle missing values, and normalize variables if necessary. Inaccurate or incomplete data can lead to biased results in spatial autocorrelation analysis.

Conclusion

Spatial autocorrelation analysis is a powerful tool for understanding the spatial patterns of variables in geospatial datasets. In this section, we discussed the concept of spatial autocorrelation, introduced the Moran's I and Geary's C statistics, and explored their interpretation. We also explored the applications of spatial autocorrelation in various fields and highlighted some practical considerations. Understanding spatial autocorrelation can provide valuable insights into spatial data analysis and guide decision-making in numerous domains.

Now, let's move on to the next section where we will delve into spatial clustering techniques.

Spatial Clustering Techniques

Spatial clustering is a fundamental concept in geospatial data analysis, which involves grouping similar geographic features together based on their spatial proximity and attribute similarities. Clustering techniques enable us to identify spatial patterns, hotspots, and outliers within a dataset, thereby gaining insights into the underlying structure and distribution of the data. In this section, we will explore some commonly used spatial clustering techniques and discuss their applications in geospatial data analysis.

K-means Clustering

One of the most widely used clustering algorithms is K-means clustering. K-means is a centroid-based clustering method that partitions a dataset into K distinct clusters, where K is a user-specified parameter. The algorithm iteratively assigns each data point to the cluster whose centroid is closest to it and then updates the centroids based on the newly assigned data points. This process continues until convergence, i.e., when the assignments no longer change significantly.

The objective function of K-means clustering is to minimize the sum of squared Euclidean distances between the data points and their respective cluster centroids. Mathematically, we can define the objective function as follows:

$$\text{minimize} \sum_{i=1}^{K} \sum_{\mathbf{x} \in C_i} ||\mathbf{x} - \boldsymbol{\mu}_i||^2$$

where C_i is the set of data points assigned to the ith cluster, $\mathbf{x}$ is a data point, and $\boldsymbol{\mu}_i$ is the centroid of the ith cluster.

K-means clustering is sensitive to the initial placement of centroids, and the final outcome of the algorithm may vary depending on the initial configuration. Therefore, it is common practice to run the algorithm multiple times with different initializations and select the best clustering solution based on a criterion such as the total within-cluster variance.

Example: K-means Clustering of Crime Data

To illustrate the application of K-means clustering, let's consider a dataset of crime incidents in a city. The dataset includes the spatial coordinates of each crime incident and additional attributes such as the type of crime and the time of occurrence.

Our goal is to identify clusters of similar crime patterns within the city. By applying K-means clustering, we can group crime incidents that share similar characteristics, such as high rates of theft or vandalism, into distinct clusters. This information can be valuable for law enforcement agencies in targeting crime prevention strategies and allocating resources effectively.

Density-based Spatial Clustering of Applications with Noise (DBSCAN)

While K-means clustering assumes that clusters are spherical-shaped and have a similar size, real-world data often exhibit complex spatial patterns and irregular cluster shapes. Density-based Spatial Clustering of Applications with Noise (DBSCAN) is a robust clustering algorithm that can detect clusters of arbitrary shape and handle noise effectively.

DBSCAN defines clusters as dense regions of data points separated by sparser regions. The algorithm starts by selecting an arbitrary data point and retrieves its neighborhood within a specified radius. If the number of data points in the neighborhood exceeds a predefined density threshold, a new cluster is formed. The process is then repeated for the newly added data points until no more points can be added to the cluster. Data points that do not meet the density criteria are labeled as noise or outliers.

DBSCAN has two key parameters: the radius (ε) defining the neighborhood of a data point and the minimum number of points ($MinPts$) required to form a dense region. These parameters control the sensitivity of the algorithm to cluster density. A smaller radius and a larger $MinPts$ value yield denser clusters, while a larger radius and a smaller $MinPts$ value allow for more sparsity in the clusters.

Example: DBSCAN for Urban Traffic Analysis

To demonstrate the utility of DBSCAN, let's consider an application in urban traffic analysis. Suppose we have a dataset of GPS trajectories from vehicles within a city. Each trajectory consists of a sequence of spatial points representing the vehicle's movement. Our objective is to identify traffic congestion patterns and detect congested areas within the city.

By applying DBSCAN to the GPS trajectory data, we can discover clusters of vehicles that share similar movement patterns, indicating congested areas. The algorithm will group trajectories of vehicles moving at slow speeds or experiencing frequent stops into clusters, while separating them from trajectories of vehicles moving smoothly in uncongested areas. This information can aid city planners and transportation authorities in identifying bottleneck locations, optimizing traffic signal timings, and devising efficient traffic management strategies.

Evaluation and Validation of Clustering Results

Evaluating and validating clustering results is essential to ensure their reliability and interpretability. Several metrics are commonly used to assess the quality of clustering solutions, including internal, external, and relative measures.

Internal measures evaluate the cohesion and separation of clusters within the same dataset. Examples of internal measures include the Silhouette coefficient, Calinski-Harabasz index, and Davies-Bouldin index. These measures assess the compactness, separation, and overall goodness of the clustering results.

External measures, on the other hand, compare the clustering results against a known ground truth or a reference dataset. This type of evaluation is applicable when the true labels or class assignments are available. Metrics such as Rand index, Adjusted Rand index, and Fowlkes-Mallows index are commonly used external measures.

In addition to internal and external measures, relative measures compare different clustering solutions based on their performance. These measures include the Dunn index, Hopkins statistic, and stability index. Relative measures help determine the optimal number of clusters and select the best clustering algorithm for a given dataset.

It is important to note that clustering is an exploratory technique, and the choice of clustering algorithm and parameters should be tailored to the specific dataset and objectives of the analysis. It is recommended to experiment with multiple clustering techniques and evaluate their results using appropriate validation measures before drawing conclusions.

Challenges and Considerations

Despite its usefulness, spatial clustering faces several challenges and considerations that need to be addressed:

- **Spatial Scale:** Clustering results can be sensitive to the choice of spatial scale or resolution. Aggregating or disaggregating the data at different scales may yield different cluster patterns, and it is important to select an appropriate scale that aligns with the underlying processes or phenomena of interest.

- **Nonstationarity:** Spatial clustering assumes stationary processes, i.e., that the statistical properties of the data are constant over space. However, many real-world phenomena exhibit nonstationarity, where the characteristics of the data vary spatially. Accounting for nonstationarity in clustering analysis is crucial to avoid spurious or biased results.

- **Outlier Detection:** Identifying and handling outliers is an important aspect of clustering analysis. Outliers can arise from measurement errors, data entry mistakes, or genuinely anomalous observations. Robust clustering algorithms, such as DBSCAN, can effectively handle outliers, but it is still important to carefully examine potential outliers and their impact on the clustering results.

- **High-dimensional Data:** Clustering high-dimensional data, such as remote sensing imagery or social media data, poses significant challenges. Curse of dimensionality, sparsity, and feature selection are some of the issues that need to be addressed when applying clustering techniques to high-dimensional datasets.

Conclusion

Spatial clustering techniques are powerful tools for uncovering hidden patterns, identifying spatial relationships, and gaining insights from geospatial datasets. K-means clustering and DBSCAN are two widely used methods that can handle different types of data and cluster structures. Evaluating clustering results using appropriate metrics and considering the challenges and considerations associated with spatial clustering are crucial for achieving meaningful and reliable results.

In the next section, we will delve into the field of geostatistical analysis, which focuses on modeling and analyzing spatially correlated data using statistical methods and techniques.

Exercises

1. Consider a dataset of housing prices in a city. Apply K-means clustering to identify clusters of houses with similar price levels. Discuss the interpretation of the clustering results and how they can guide real estate decision-making.

2. Imagine you have a dataset of customer locations and purchase histories for an online retail company. Apply DBSCAN to identify clusters of frequent customers and outliers representing irregular purchase behavior. Discuss how these clusters can help improve marketing strategies and customer targeting.

3. Investigate the sensitivity of K-means clustering to different initializations. Randomly initialize the centroids in multiple runs of the algorithm and compare the resulting clustering solutions using a suitable evaluation metric.

4. Explore the impact of varying the radius (ε) and minimum points ($MinPts$) parameters in DBSCAN on clustering results. Discuss how different values of these parameters affect the cluster density and the identification of noise points.

Note: Solutions to the exercises can be found in the "Solutions" section at the end of the book.

Additional Resources

To further enhance your understanding of spatial clustering techniques, consider exploring the following resources:

- Books:
  - Anselin, L., Rey, S. J., & Getis, A. (2014). *Spatial Analysis: Statistics, Visualization, and Computational Methods*. SAGE Publications.
  - Fischer, M. M., & Getis, A. (Eds.). (2010). *Handbook of Applied Spatial Analysis: Software Tools, Methods and Applications*. Springer.
  - Hahsler, M., Piekenbrock, M., & Dorau, R. (2021). *Introduction to Data Science: Data Analysis and Prediction Algorithms with R*. Springer.
- Articles:
  - Arthur, D., & Vassilvitskii, S. (2007). *k-means++: The advantages of careful seeding*. Proceedings of the Eighteenth Annual ACM-SIAM Symposium on Discrete Algorithms.
  - Ester, M., Kriegel, H. P., Sander, J., & Xu, X. (1996). *A density-based algorithm for discovering clusters in large spatial databases with noise*. Proceedings of the Second International Conference on Knowledge Discovery and Data Mining.
  - Huang, H., & Qian, Z. (2015). *Comparing spatial clustering methods for exploratory point data analysis: a case study of crime incidents in Wuhan, China*. GeoJournal, 80(1), 57-74.
- Software Packages:

- Scikit-learn: a Python library for machine learning, which provides implementations of K-means clustering and DBSCAN algorithms. Documentation available at: `https://scikit-learn.org/stable/`
- PySAL: a Python library for spatial analysis, which includes various clustering algorithms and evaluation metrics. Documentation available at: `https://pysal.org/`

Solutions to Exercises

Exercise 1

Applying K-means clustering to the dataset of housing prices can reveal clusters of houses with similar price levels. The clustering results provide insights into the spatial distribution of housing prices, allowing real estate agents or investors to make informed decisions.

Interpretation of clustering results can involve examining the characteristics of each cluster, such as the average price, median price, or price range. For example, one cluster may consist of high-end luxury houses with above-average prices, while another cluster may include affordable houses with below-average prices. Analyzing the spatial patterns of these clusters can uncover valuable information, such as the presence of upscale neighborhoods or areas with more affordable housing options. This information can guide real estate professionals in identifying target markets, setting pricing strategies, and determining investment opportunities.

Exercise 2

DBSCAN applied to the dataset of customer locations and purchase histories can identify clusters of frequent customers, representing loyal and valuable customer segments. Outliers or noise points in the dataset can also be detected, signifying irregular purchase behavior.

Analyzing these clusters can help improve marketing strategies and customer targeting. For instance, marketing campaigns can be tailored specifically to the different cluster groups to increase customer engagement and satisfaction. Cluster analysis can also assist in identifying patterns in customer preferences and behavior, allowing for personalized product recommendations or targeted offers. Additionally, the detection of outliers can uncover potential customer dissatisfaction or unusual purchasing patterns that may warrant further investigation.

Exercise 3

To investigate the sensitivity of K-means clustering to different initializations, we can run the algorithm multiple times with varying initial centroid positions. For each run, we can calculate the total within-cluster variance as an evaluation metric. The lower the within-cluster variance, the better the clustering solution.

Here's a step-by-step approach to addressing this exercise:

1. Randomly initialize the centroids for K-means clustering.

2. Run the algorithm and calculate the total within-cluster variance.

3. Repeat steps 1 and 2 multiple times (e.g., 10 times) with different initializations.

4. Compare the clustering solutions based on the total within-cluster variance.

5. Select the solution with the lowest total within-cluster variance as the best clustering solution.

By comparing the clustering solutions, you can gain insights into the stability and robustness of the K-means algorithm. Different initializations may yield slightly different results, indicating the presence of multiple local optima. It is important to consider this variability when interpreting and trusting the clustering outcome.

Exercise 4

To explore the impact of varying the radius (ε) and minimum points ($MinPts$) parameters in DBSCAN, we can perform sensitivity analysis by running the algorithm with different parameter settings. By comparing the resulting clustering solutions, we can assess how changes in these parameters affect the cluster density and the identification of noise points.

Here's a step-by-step approach to addressing this exercise:

1. Select multiple values for the radius (ε) parameter, ranging from small to large values.
2. Select multiple values for the minimum points ($MinPts$) parameter, ranging from small to large values.
3. Run DBSCAN using each combination of the parameter values.
4. Compare the clustering solutions based on cluster density and the number of noise points.
5. Analyze the differences in the clustering results for each parameter combination and draw conclusions about the impact of the radius and minimum points parameters.

By conducting this analysis, you can determine the optimal parameter values for detecting clusters of desired density and effectively filtering out noise points. Additionally, you can gain insights into the sensitivity of DBSCAN to these parameter settings and understand the trade-offs between cluster density, cluster shape, and the presence of noise points.

Geospatial Data Mining Techniques

In this section, we will explore the exciting field of geospatial data mining. Geospatial data mining refers to the process of discovering patterns, relationships, and insights from spatially explicit data. It involves applying various data mining techniques to analyze and extract meaningful information from geospatial datasets. Geospatial data mining plays a crucial role in many real-world applications, such as urban planning, environmental management, public health, and natural hazards analysis.

Introduction to Geospatial Data Mining

Geospatial data mining combines concepts from data mining, spatial analysis, and geographic information systems (GIS) to uncover hidden patterns and trends within spatial datasets. It involves the extraction of useful knowledge from large volumes of geospatial data. Geospatial data mining techniques can be divided into supervised and unsupervised methods.

In supervised geospatial data mining, we have a target variable that we aim to predict or classify based on a set of input features. This approach is useful when we have labeled or categorized observation data. On the other hand, unsupervised geospatial data mining techniques do not require prior knowledge of the target variable. Instead, they aim to discover inherent structures or patterns within the dataset.

Spatial Association Rules Mining

Spatial association rules mining is a popular technique in geospatial data mining. It aims to discover interesting relationships or associations between spatial features. The output of this technique is a set of association rules that capture the spatial dependencies between different geographic objects.

To illustrate this technique, imagine we have a dataset of crime incidents in a city, with each incident associated with a specific location. By applying spatial association rules mining, we can identify interesting relationships between different types of crimes and specific locations. For example, we may discover that certain types of crime tend to occur more frequently in proximity to particular amenities, such as bars or parks. This information can be valuable for urban planning, crime prevention, and resource allocation.

Cluster Analysis

Cluster analysis is another powerful geospatial data mining technique. It involves grouping similar spatial objects into clusters based on their spatial proximity or other similarity measures. Cluster analysis helps identify spatial patterns and structures within geospatial datasets.

For example, in urban planning, cluster analysis can be used to identify areas with similar characteristics, such as high population density, income levels, or land-use patterns. This information can inform decisions regarding infrastructure development, resource allocation, and zoning regulations.

Various clustering algorithms can be used in geospatial data mining, including K-means, DBSCAN (Density-Based Spatial Clustering of Applications with Noise), and hierarchical clustering. Each algorithm has its advantages and limitations, and the choice of algorithm depends on the specific application and dataset.

Spatial Outlier Detection

Outlier detection is an essential step in geospatial data mining. Outliers are data points that deviate significantly from the expected or typical pattern. Detecting spatial outliers can help identify unusual or anomalous spatial phenomena.

For instance, in environmental management, spatial outlier detection can be used to identify areas with exceptionally high pollution levels or detect unusual patterns in species distribution.

By identifying these outliers, appropriate actions can be taken, such as targeted pollution control measures or conservation efforts.

There are several approaches to spatial outlier detection, including statistical methods, distance-based methods, and local outlier factor analysis. These methods leverage spatial attributes, such as spatial autocorrelation or distance to neighbors, to identify outliers.

Predictive Modeling

Predictive modeling is a fundamental technique in geospatial data mining. It involves training a predictive model that can accurately predict a target variable based on input features. Predictive modeling is widely used in various domains, such as transportation planning, public health, and natural resource management.

For example, in public health, predictive modeling can be used to forecast the spread of infectious diseases based on various factors such as population density, socio-economic indicators, and environmental factors. This information can help public health officials allocate resources, implement preventive measures, and plan for healthcare service delivery.

Several machine learning algorithms can be employed for predictive modeling in geospatial data mining, including decision trees, random forests, support vector machines, and neural networks. The choice of algorithm depends on the nature of the data, the target variable, and the desired prediction accuracy.

Geospatial Data Mining Challenges and Future Directions

While geospatial data mining offers immense potential, it also presents several challenges. Some of the key challenges include data heterogeneity, data quality, computational complexity, and privacy concerns. Addressing these challenges requires interdisciplinary collaborations and the development of advanced data mining techniques.

In the future, advancements in data mining algorithms, distributed computing, and machine learning will continue to shape the field of geospatial data mining. Moreover, the integration of emerging technologies such as remote sensing, Internet of Things (IoT), and big data analytics will further enhance our ability to extract valuable insights from geospatial data.

Conclusion

Geospatial data mining techniques provide powerful tools for analyzing and extracting knowledge from spatial datasets. Whether it is discovering spatial associations, identifying clusters, detecting outliers, or building predictive models, geospatial data mining plays a vital role in decision-making processes across various domains. By leveraging the vast amounts of geospatial information available today, we can gain a deeper understanding of the world around us and make informed decisions to address complex spatial challenges.

Further Reading

For further reading on geospatial data mining, the following resources are recommended:

- Shekhar, S., Huang, Y., & Zhou, X. (Eds.). (2011). *Geospatial data mining and knowledge discovery*. CRC Press.
- Miller, H. J., & Han, J. (Eds.). (2009). *Geographic data mining and knowledge discovery*. CRC Press.

- Longley, P. A., Goodchild, M. F., Maguire, D. J., & Rhind, D. W. (2015). *Geographic information science & systems.* John Wiley & Sons.

These resources provide in-depth coverage of geospatial data mining techniques, algorithms, and applications, and can serve as valuable references for further exploration in the field.

Geostatistical Analysis

Introduction to Geostatistics

Geostatistics is a field of study that deals with the analysis and interpretation of spatial data. It provides a set of powerful statistical techniques to model and analyze spatial patterns, enabling us to make predictions in unobserved locations. Geostatistics has widespread applications in various disciplines, including geology, environmental science, agriculture, and urban planning.

In this section, we will introduce the fundamental principles of geostatistics and discuss its relevance in geospatial data analysis. We will explore the key concepts, methods, and models used in geostatistics, along with practical examples and exercises to enhance your understanding.

Spatial Autocorrelation

One of the fundamental concepts in geostatistics is spatial autocorrelation. It refers to the tendency of spatially close observations to have similar values or attributes. Spatial autocorrelation can be positive, indicating that similar values cluster together, or negative, indicating that dissimilar values tend to be close to each other.

Understanding spatial autocorrelation is crucial in geostatistics as it allows us to explore the underlying spatial structure of the data. By quantifying the degree of spatial autocorrelation, we can determine the spatial scale at which patterns exist, identify spatial clusters, and assess the level of spatial dependence in our data.

Variograms

To quantify spatial autocorrelation, we use variograms. A variogram measures the degree of dissimilarity between pairs of observations as a function of their spatial separation or lag. It provides important information about the spatial structure of the data and helps us select appropriate geostatistical models.

The variogram is calculated by taking pairs of observations at different lags and computing the variance of their differences. The resulting variogram function is typically graphed as a plot of the variance against the lag distance. The shape of the variogram curve provides insights into the spatial dependence structure of the data, such as the range, sill, and nugget effect.

The range represents the distance at which spatial autocorrelation becomes negligible, indicating the maximum distance at which one observation has an impact on another. The sill is the limiting value of the variogram at larger lags, representing the maximum variability in the data. The nugget effect is a discontinuity at the origin of the variogram, indicating the presence of spatial variability at extremely short distances.

Kriging

Kriging is a widely used geostatistical interpolation technique that provides optimal predictions of values at unobserved locations based on observed data. It takes into account both the spatial autocorrelation and the uncertainty associated with the predictions.

There are different types of kriging methods, including ordinary kriging, universal kriging, and co-kriging. Ordinary kriging assumes a constant mean and estimates the spatial autocorrelation structure from the data. Universal kriging incorporates additional external variables to account for spatially varying trends. Co-kriging uses auxiliary variables to improve the predictions, especially when there are limited observations for the target variable.

By estimating the spatial autocorrelation structure from the variogram, kriging provides predictions with the best possible accuracy at unsampled locations. It also generates prediction standard errors, allowing us to quantify the uncertainty associated with the predictions.

Model Selection and Validation

Selecting an appropriate geostatistical model is crucial for accurate predictions. Improper model selection can lead to biased and unreliable results. To select the best model, we compare different models using goodness-of-fit statistics, such as the mean squared error or the Akaike information criterion (AIC).

Model validation is an essential step in geostatistics to assess the performance of the selected model and determine its predictive accuracy. Cross-validation techniques, such as k-fold cross-validation or leave-one-out cross-validation, can be used to validate the model performance by comparing the predicted values with the observed values at validation locations.

Practical Application

Geostatistics finds applications in various fields, including natural resource management, environmental monitoring, and urban planning. For example, geostatistics can be used to model and predict soil contamination levels in an area based on a limited number of soil samples. It can also help analyze the spread of air pollutants in a city and predict future contamination levels.

In the next section, we will dive deeper into the specific techniques used in geostatistics, such as ordinary kriging, co-kriging, and Bayesian geostatistics. We will explore their algorithms, assumptions, and practical examples to illustrate their applications in geospatial data analysis.

Summary

In this section, we provided an introduction to geostatistics and its relevance in geospatial data analysis. We discussed spatial autocorrelation as a fundamental concept in geostatistics and how it helps identify spatial patterns and clusters. We introduced variograms as a tool to quantify spatial autocorrelation and understand the spatial structure of the data. Kriging, a widely used geostatistical interpolation technique, was introduced as a method to provide optimal predictions at unobserved locations. We also discussed the importance of model selection and validation in geostatistics. Finally, we explored the practical applications of geostatistics in various fields.

In the next section, we will delve deeper into geostatistical analysis by exploring the techniques of ordinary kriging, co-kriging, and Bayesian geostatistics. We will provide detailed explanations, examples, and exercises to enhance your understanding of these advanced geostatistical methods.

Ordinary Kriging

In geostatistics, Ordinary Kriging is a spatial interpolation technique used to estimate values at unsampled locations based on the surrounding sampled data points. It is widely used in geospatial data analysis to create continuous surfaces or maps from discrete point data. Ordinary Kriging takes into account the spatial correlation between data points and generates unbiased estimates with minimum error variance.

Background

Kriging, named after the South African mining engineer Danie G. Krige, is a geostatistical technique developed in the 1950s. It is a powerful tool for analyzing spatially correlated data and has applications in various fields such as remote sensing, environmental monitoring, and resource management.

The fundamental principle of Kriging is based on the spatial autocorrelation of data, which assumes that data points that are closer together are more similar than those far apart. Ordinary Kriging builds upon this principle by estimating unknown values at unsampled locations as a linear combination of nearby data points, where the weights are determined based on the spatial autocorrelation structure of the data.

Theory of Ordinary Kriging

Let's consider a set of observed data points in a two-dimensional space. Each data point has a known value and a corresponding spatial location. We want to estimate the value at an unobserved location within the study area.

To perform Ordinary Kriging, we assume that the data follows a stationary and isotropic spatial process. Stationarity means that the statistical properties of the data do not change over space, while isotropy assumes that the spatial correlation depends only on the distance between points and not on their orientation.

The Ordinary Kriging estimation at an unsampled location can be expressed as:

$$Z(u_0) = \sum_{i=1}^{n} \lambda_i Z(u_i)$$

where $Z(u_0)$ is the estimated value at the unsampled location, $Z(u_i)$ are the observed values at the sampled locations, λ_i are the weights assigned to each observed value, and n is the total number of observed data points.

The weights, or Kriging coefficients, are determined based on the spatial autocovariance function, also known as the variogram. The variogram measures the degree of spatial dependence between data points as a function of their separation distance. By fitting a model to the variogram, we obtain information about the spatial correlation structure of the data.

The Kriging coefficients can be obtained by solving the system of linear equations derived from the variogram model and the spatial dependence between the observed and unsampled locations. The coefficients are calculated to minimize the error variance between the estimated values and the observed values.

Steps in Ordinary Kriging

The process of performing Ordinary Kriging involves the following steps:

1. Data Preparation: Collect the required spatial data and preprocess it to ensure data quality.

2. Variogram Estimation: Calculate the empirical variogram to investigate the spatial dependence structure of the data.

3. Variogram Modeling: Fit a mathematical model to the empirical variogram to capture the spatial correlation.

4. Kriging System Setup: Compute the weights (Kriging coefficients) based on the variogram model and solve the system of linear equations.

5. Kriging Estimation: Use the calculated weights to estimate values at unsampled locations.

6. Error Estimation: Assess the uncertainty of the Kriging estimates using measures such as the Kriging variance.

Example

Let's consider a hypothetical scenario where we have a set of soil contamination data points in a contaminated site. We want to estimate the soil contamination levels at unsampled locations within the site.

We start by collecting soil samples at various locations within the site and measure their contamination levels. We then preprocess the data to remove any outliers and ensure its quality.

Next, we calculate the empirical variogram by pairwise comparing the contamination level differences between the sampled locations. The variogram shows that the spatial correlation decreases with increasing distance between points.

We fit a mathematical model, such as the spherical, exponential, or Gaussian model, to the empirical variogram to describe the spatial correlation accurately.

Using the variogram model, we compute the Kriging weights by solving the system of linear equations derived from the variogram. These weights represent how much influence each observed data point has on the estimation at unsampled locations.

Finally, we use the calculated weights to estimate the soil contamination levels at the unsampled locations within the contaminated site. We also quantify the uncertainty associated with the estimates using the Kriging variance.

Resources and Further Reading

To learn more about Ordinary Kriging and its application in geospatial data analysis, the following resources are recommended:

- Cressie, N.A.C. Statistics for Spatial Data (Revised Edition). Wiley, 1993.
- Deutsch, C.V., & Journel, A.G. GSLIB: Geostatistical Software Library and User's Guide (2nd Edition). Oxford University Press, 1998.
- Kang-tsung Chang. Introduction to Geographic Information Systems (8th Edition). McGraw-Hill Education, 2015.
- PySAL: Python Spatial Analysis Library. `https://pysal.org/`

These resources provide comprehensive coverage of geostatistical techniques, including Ordinary Kriging, and offer practical examples and case studies for deeper understanding.

Conclusion

Ordinary Kriging is a powerful geostatistical method used for spatial interpolation and estimation of values at unsampled locations. By considering the spatial correlation between data points, Ordinary Kriging generates unbiased estimates with minimum error variance. Understanding the theory and application of Ordinary Kriging is essential for anyone interested in geospatial data analysis and mapping with Python.

Co-Kriging and Co-Regression

In geostatistics, Co-Kriging and Co-Regression are two closely related techniques used for estimating and predicting values at unsampled locations based on the relationships with other observed variables. These techniques are widely used in spatial data analysis and can provide valuable insights into the spatial variability of a phenomenon.

Introduction to Co-Kriging

Co-Kriging is an extension of the Ordinary Kriging technique that incorporates additional auxiliary variables to improve the accuracy of predictions. In Ordinary Kriging, the estimation is based solely on the correlation structure of the variable of interest. However, in Co-Kriging, the correlation between the variable of interest and other related variables is also taken into account.

The basic idea behind Co-Kriging is to exploit the relationship between the primary variable and the auxiliary variables to improve the estimation accuracy. By including additional information from correlated variables, Co-Kriging can reduce the uncertainty in the predictions and provide more reliable results.

Co-Kriging Procedure

The Co-Kriging procedure involves several steps:

1. Data Collection: Collect the primary variable and auxiliary variables at each location of interest. Ensure that the variables are measured on the same scale and have a similar spatial support.

2. Exploratory Data Analysis: Perform exploratory data analysis to understand the spatial relationships between the primary variable and the auxiliary variables. This can include techniques such as scatter plots, correlation analysis, and variogram analysis.

3. Variogram Analysis: Calculate the variograms for the primary variable and the auxiliary variables. The variogram describes the spatial structure and variability of the data.

4. Model Fitting: Fit variogram models to the experimental variograms of the primary variable and the auxiliary variables. The variogram models characterize the spatial dependence and provide the parameters required for Kriging.

5. Co-Kriging Estimation: Estimate the values at unsampled locations using Co-Kriging. This involves calculating the covariance between the primary variable and the auxiliary variables and using the estimated variogram models to interpolate the values.

6. Prediction Accuracy Assessment: Assess the accuracy of the Co-Kriging predictions using cross-validation or other validation techniques. Compare the predicted values with the observed values to evaluate the performance of the Co-Kriging model.

Co-Regression

Co-Regression is a similar technique to Co-Kriging, but it focuses on the relationship between the primary variable and the auxiliary variables using regression models. In Co-Regression, instead of modeling the spatial correlation using variograms, the relationship between the variables is captured using regression equations.

The Co-Regression model estimates the relationship between the primary variable and the auxiliary variables at each location. This relationship is then used to predict the values at unsampled locations based on the observed values of the auxiliary variables. The Co-Regression model can be linear or non-linear, depending on the nature of the relationship between the variables.

Co-Regression can be a useful technique when the spatial dependence between the variables is not well captured by the variogram models or when the auxiliary variables have a strong predictive power for the primary variable.

Advantages of Co-Kriging and Co-Regression

Co-Kriging and Co-Regression offer several advantages over traditional interpolation and prediction techniques:

1. Incorporation of Auxiliary Information: By including additional information from correlated variables, Co-Kriging and Co-Regression can improve the accuracy of predictions and reduce the uncertainty associated with the estimates.
2. Explicit Modeling of Spatial Dependence: Co-Kriging and Co-Regression explicitly model the spatial dependence between the variables, taking into account both the correlation and the spatial structure of the data.
3. Handling of Missing Data: Co-Kriging and Co-Regression can handle missing data by using the available measurements from the auxiliary variables to predict the missing values in the primary variable.
4. Realistic Predictions: Co-Kriging and Co-Regression can provide realistic predictions by incorporating the spatial information and the relationships between the variables.

Limitations and Challenges

While Co-Kriging and Co-Regression have many advantages, there are also some limitations and challenges associated with their implementation:

1. Assumptions: Co-Kriging and Co-Regression assume a linear relationship between the primary variable and the auxiliary variables. If the relationship is non-linear, alternative regression techniques or machine learning algorithms may be more appropriate.

2. Spatial Homogeneity: Co-Kriging and Co-Regression assume spatial homogeneity, which means that the relationship between the variables does not vary across space. If the relationship varies spatially, local models or spatially-varying regression techniques may be required.

3. Selection of Auxiliary Variables: The selection of appropriate auxiliary variables is crucial for the success of Co-Kriging and Co-Regression. Care should be taken to choose variables that are correlated with the primary variable and provide additional information.

4. Data Quality and Availability: Co-Kriging and Co-Regression require high-quality data that is collected at a sufficient number of locations. Sparse or low-quality data can lead to inaccurate predictions and unreliable results.

Case Study: Soil Nutrient Mapping

To provide a practical example of Co-Kriging and Co-Regression, let's consider the case of soil nutrient mapping. Soil nutrients are essential for agriculture and understanding their spatial distribution is crucial for optimizing fertilizer applications.

In this case study, we have collected data on soil nutrients such as nitrogen, phosphorus, and potassium at various locations within a farming region. Additionally, we have collected data on several auxiliary variables, including elevation, slope, and land use.

Using Co-Kriging, we can estimate the soil nutrient levels at unsampled locations by incorporating the relationships between the primary variables (nitrogen, phosphorus, and potassium) and the auxiliary variables (elevation, slope, and land use). The Co-Kriging predictions will provide a more accurate representation of the spatial distribution of soil nutrient levels compared to Ordinary Kriging.

Alternatively, Co-Regression can be employed to model the dependencies between the soil nutrient levels and the auxiliary variables using regression models. The Co-Regression models can then be used to predict the soil nutrient levels at unsampled locations based on the observed values of the auxiliary variables.

Both Co-Kriging and Co-Regression can provide valuable insights into the spatial patterns and variability of soil nutrient levels, enabling farmers to make informed decisions regarding fertilizer applications and soil management practices.

Conclusion

Co-Kriging and Co-Regression are powerful techniques in geospatial data analysis that allow us to incorporate auxiliary information to improve the accuracy of predictions. By explicitly modeling the spatial relationships between variables, these techniques provide valuable insights into the spatial variability of phenomena.

In this section, we introduced the concepts of Co-Kriging and Co-Regression, discussed their procedures, advantages, and limitations, and provided a case study on soil nutrient mapping. These techniques can be applied to various fields, including environmental studies, urban planning, and natural resource management, to analyze and predict spatial phenomena accurately.

Bayesian Geostatistics

In this section, we will explore the concept and application of Bayesian geostatistics in geospatial data analysis. Geostatistics is a field that focuses on analyzing and modeling spatial phenomena based on data collected at specific locations. Bayesian geostatistics combines geostatistical techniques with Bayesian inference to provide a flexible and robust framework for spatial data analysis.

Background of Bayesian Inference

Before delving into Bayesian geostatistics, let's briefly discuss the principles of Bayesian inference. Bayesian inference is a statistical approach that allows for the updating of prior beliefs or knowledge about a parameter or a hypothesis based on observed data. It provides a way to quantify uncertainty and make probabilistic predictions.

Bayesian inference involves specifying prior distributions, likelihood functions, and posterior distributions. The prior distribution represents the existing knowledge or beliefs about the parameter before observing any data. The likelihood function describes the probability of observing the data given the parameter values. The posterior distribution, which is proportional to the product of the prior and the likelihood, represents the updated knowledge about the parameter after considering the observed data.

Bayesian Geostatistical Models

Bayesian geostatistical models are used to analyze and model spatial data in a probabilistic framework. These models take into account both the spatial dependence between locations and the uncertainty associated with the estimations. Compared to traditional geostatistical models, Bayesian geostatistical models offer several advantages, including the ability to incorporate prior information, handle missing data, and provide probabilistic predictions.

A common approach in Bayesian geostatistics is to use Gaussian processes as priors for spatial random fields. Gaussian processes are flexible and can capture various spatial patterns. The spatial dependence structure is characterized by a covariance function, which describes the correlation between spatial locations. In Bayesian geostatistics, the parameters of the covariance function can be assigned prior distributions, allowing for uncertainty quantification.

Bayesian Inference in Geostatistics

Performing Bayesian inference in geostatistics involves estimating the posterior distribution of the model parameters given the observed data. This estimation is often achieved through Markov chain Monte Carlo (MCMC) methods, such as the Metropolis-Hastings algorithm or the Gibbs sampler. These algorithms allow for sampling from the posterior distribution even when it is analytically intractable.

The MCMC methods iteratively sample from the parameter space, guided by the posterior distribution. The samples obtained from the MCMC chain provide information about the posterior distribution, and, subsequently, about the uncertainty in the estimated parameters and predictions. By summarizing the MCMC samples, one can obtain point estimates, credible intervals, and other statistical summaries.

Model Selection and Comparison

In Bayesian geostatistics, model selection and comparison are essential steps in the analysis. Model selection refers to choosing the most appropriate model among a set of candidate models based on the observed data. Model comparison, on the other hand, involves assessing the relative performance of different models.

Bayesian model selection and comparison can be performed using various techniques, such as the Deviance Information Criterion (DIC) and the Bayesian Information Criterion (BIC). These criteria balance the goodness of fit of the model with the complexity of the model.

Case Study: Spatial Interpolation of Air Quality Data

To illustrate the application of Bayesian geostatistics, let's consider a case study involving the spatial interpolation of air quality data. Air quality is an important environmental parameter that affects public health and urban planning decisions. In this case study, we aim to estimate the concentrations of a specific pollutant across a region of interest.

The first step is to collect air quality measurements at a set of monitoring stations scattered across the region. We can then use these measurements to model the spatial distribution of the pollutant using a Bayesian geostatistical model. By incorporating information about the spatial dependence and uncertainty, we can produce maps that depict the estimated pollutant concentrations with associated uncertainty.

The Bayesian geostatistical model will involve specifying prior distributions for the model parameters, such as the spatial covariance parameters and the measurement error variance. The model will also include a likelihood function that describes the probability of observing the air quality measurements given the spatial distribution. By combining the prior and likelihood, we can use MCMC methods to estimate the posterior distribution of the model parameters.

Once the model parameters are estimated, we can generate spatial predictions of the pollutant concentrations at unobserved locations. These predictions will not only provide point estimates but also quantify the uncertainty associated with the predictions. This information can be used to inform decision-making processes and assess the potential risk related to air pollution exposure.

Conclusion

Bayesian geostatistics offers a powerful framework for analyzing and modeling spatial data. By incorporating prior information and providing probabilistic predictions, Bayesian geostatistical models allow for a more comprehensive understanding of spatial phenomena. Through the use of MCMC methods, uncertainty can be quantified, and model selection and comparison can be performed. The application of Bayesian geostatistics in geospatial data analysis opens up new possibilities for addressing contemporary challenges in various domains, such as urban planning, environmental management, and public health. By embracing these techniques, researchers and practitioners can gain deeper insights into spatial processes and make informed decisions based on robust and rigorous analyses.

Resources and Further Reading

If you are interested in learning more about Bayesian geostatistics, here are some recommended resources:

- Cressie, N., & Wikle, C. K. (2015). *Statistics for Spatio-Temporal Data*. John Wiley & Sons.
- Diggle, P., Tawn, J., & Moyeed, R. (1998). *Model-based geostatistics*. Journal of the Royal Statistical Society: Series C (Applied Statistics), 47(3), 299-350.
- Law, J., & Illian, J. (2018). *Bayesian Methods in Environmental Science*. CRC Press.

These resources provide more in-depth coverage of Bayesian geostatistics, including theory, methodologies, and practical examples. Additionally, academic journals in the field of spatial statistics and environmental sciences often publish research articles on Bayesian geostatistics, offering the latest advancements and applications in the field.

Spatial Data Visualization and Cartography

Principles of Data Visualization

Data visualization is a crucial aspect of geospatial data analysis. It allows us to explore and understand complex datasets by representing information visually. In this section, we will discuss the principles of data visualization and the importance of designing effective and informative visualizations.

The Importance of Visualization

Data visualization is an essential tool in geospatial data analysis as it helps us communicate insights and patterns effectively. By transforming raw data into visual representations, we can easily identify trends, outliers, and relationships, which may not be apparent in raw data alone. Visualization enables us to gain a deeper understanding of the underlying patterns, helping us make informed decisions and communicate information to various stakeholders.

Visual Perception and Cognition

To design effective visualizations, it is crucial to understand how humans perceive and interpret visual information. Human perception and cognition play a significant role in data visualization, as it determines how well people can understand and interpret visual representations.

Certain visual properties, such as color, shape, size, and position, are more easily perceived by humans. For example, we perceive larger objects as being more significant and brighter colors as more intense. Understanding these perceptual properties can help us design visualizations that effectively convey information.

Visual Encoding

Visual encoding refers to the mapping of data attributes to visual properties. By carefully selecting appropriate visual encodings, we can effectively represent data in a visual form. Different types of data (e.g., categorical, numerical) and the relationships between them require different visual encodings.

Common visual encodings include:

- **Position:** The spatial location of objects can be used to represent data. For example, a scatter plot uses the x and y positions of points to show the relationship between two variables.

- **Color:** Color can be used to represent categorical or continuous data. For categorical data, different colors can be assigned to different categories. For continuous data, a gradient of colors can be used to represent varying values.

- **Size:** The size of objects can be used to represent numerical values. For example, a bubble chart uses the size of bubbles to represent a third variable in addition to the x and y positions.

- **Shape:** Different shapes can be used to represent different categories or groups. This encoding is particularly useful when color may not be distinguishable due to colorblindness or printing limitations.

- **Texture:** Patterns or textures can be used to represent different categories or densities. However, it is important to use texture sparingly as it can introduce unnecessary visual clutter.

The choice of visual encoding depends on the type and characteristics of the data, as well as the goals of the visualization.

Choosing the Right Visualization

Selecting the appropriate visualization technique is paramount to effectively communicate the intended message and insight. It requires considering the characteristics of the data, the relationships between variables, and the objectives of the analysis.

Here are some common types of visualizations used in geospatial data analysis:

- **Choropleth Map:** A choropleth map uses different colors or patterns to represent different values or categories within predefined geographic areas. It is commonly used to display population density, election results, or any other data that is related to specific regions.

- **Scatter Plot:** Scatter plots use the x and y positions of points to represent the relationship between two variables. They are useful for identifying correlations, outliers, and clusters.

- **Heatmap:** A heatmap visualizes the density or intensity of a variable across a geographic area. It uses color gradients to represent varying values, allowing the viewer to identify hotspots or cold spots.
- **Network Diagram:** Network diagrams represent the relationships between different entities or nodes. They are useful for visualizing social networks, transportation networks, or any other interconnected systems.
- **Time Series Plot:** Time series plots show the changes in a variable over time. They are commonly used to analyze trends and patterns in temporal data.

The choice of visualization technique should consider the nature of the data and the insights that need to be communicated.

Design Principles

To create effective and compelling visualizations, it is important to follow design principles that enhance clarity and understanding. Some key principles include:

- **Simplicity:** Keep the visual design simple and uncluttered to focus the viewer's attention on the essential elements. Remove any unnecessary elements that do not contribute to the message.
- **Consistency:** Use consistent visual encoding across different parts of the visualization to avoid confusion and facilitate comparison. For example, use the same color scheme to represent the same categories throughout the visualization.
- **Hierarchy:** Organize the visual elements in a logical and hierarchical manner to guide the viewer's attention.

Establish clear visual hierarchies to emphasize important information and de-emphasize less relevant details.

- **Accessibility:** Consider the needs of various users, including those with colorblindness or visual impairments. Choose colors and design elements that are accessible and consider providing alternative text descriptions for visual elements.

By applying these design principles, we can create visualizations that are more effective in conveying information and insights.

Tips and Tricks

Here are some additional tips and tricks for creating effective visualizations:

- **Know your audience:** Understand the background, knowledge, and goals of your audience. This will help you tailor the visualization to their specific needs and make it more relatable.

- **Tell a story:** Use the visualization to tell a coherent and compelling story. Structure the visualization in a way that guides the viewer through the narrative, highlighting the key insights and takeaways.

- **Iterate and seek feedback:** Visualizations often require multiple iterations to refine and improve. Seek feedback from colleagues or domain experts to identify areas of improvement and make necessary adjustments.

- **Experiment with interactivity:** Utilize interactive elements, such as tooltips or interactive filters, to allow users to explore the data and gain deeper insights.

Interactivity can provide a more engaging and dynamic experience.

By applying these tips and tricks, we can create visualizations that are not only informative but also engaging and impactful.

Conclusion

In this section, we have explored the principles of data visualization for geospatial data analysis. We discussed the importance of visualization, visual perception and cognition, visual encoding, choosing the right visualization technique, design principles, and provided tips and tricks for creating effective visualizations.

By following these principles and considering the specific characteristics of the data and the objectives of the analysis, we can create visualizations that effectively communicate insights, facilitate understanding, and support decision-making in geospatial data analysis. Visualization is a powerful tool that allows us to unlock the potential of our data and communicate complex concepts in a visual and intuitive manner.

Mapping with Python Libraries

In this section, we will explore the various Python libraries available for mapping and visualize geospatial data. These libraries provide a wide range of tools and techniques for creating maps, customizing their appearance, and overlaying additional layers of information. We will discuss some of the most popular libraries, including Matplotlib, Basemap, and Cartopy.

Background

Mapping is an essential aspect of geospatial data analysis, allowing us to represent complex geographical information in a visual and intuitive manner. Mapping can reveal patterns, relationships, and trends that may not be evident from raw data alone. Python libraries offer powerful tools for data visualization and mapping, making it easier to create informative and visually appealing maps.

Matplotlib

Matplotlib is a widely used plotting library in the Python ecosystem that can be used for creating basic maps. It provides a versatile set of functions for creating different types of plots, including line plots, scatter plots, bar plots, and more. Although Matplotlib is not specifically designed for mapping, it can be used effectively for simple map visualizations.

To create a basic map using Matplotlib, we need to have a set of coordinates that define the boundary of the area we want to map. We can then use the Matplotlib function `plot()` to plot these coordinates as a polygon, representing the boundary of the area. We can further customize the appearance of the map by adding labels, legends, and other elements using the various functions available in Matplotlib.

Basemap

Basemap is a Python library built on top of Matplotlib that provides additional functionalities specifically designed for creating maps. It provides a comprehensive set of map projections, which allow us to represent the surface of the Earth on a two-dimensional plane with minimal distortion. Basemap also includes built-in support for geospatial data formats, such as shapefiles, making it easier to work with a wide range of geospatial data sources.

To create a map using Basemap, we first need to initialize a map with a specific projection. We can then plot various map elements, such as coastlines, rivers, and political boundaries, using the functions provided by Basemap. We can also overlay additional layers of information, such as points, lines, and polygons representing different geospatial features. Basemap offers extensive customization options, allowing us to control the appearance of the map elements, including colors, line styles, and labels.

Cartopy

Cartopy is another popular mapping library in the Python ecosystem, which combines the capabilities of Basemap with a more modern, object-oriented interface. It provides a flexible and intuitive API for creating maps and working with geospatial data. Cartopy also supports a wide range of map projections and provides seamless integration with Matplotlib, making it easier to create publication-quality maps.

To create a map using Cartopy, we first need to define the desired map projection and initialize a new map object. We can then add various map elements, such as coastlines, rivers, and political boundaries, using the available functions in Cartopy. Similar to Basemap, Cartopy supports the overlaying of additional layers of information and provides extensive customization options for map appearance.

Comparison and Recommendation

While all three libraries (Matplotlib, Basemap, and Cartopy) offer capabilities for creating maps, they have different strengths and weaknesses. Matplotlib is the most basic library for plotting and mapping, suitable for simple map visualizations. Basemap provides more advanced functionalities specifically designed for mapping, such as map projections and geospatial data support. Cartopy builds upon Basemap and offers a more modern and flexible interface with seamless integration with Matplotlib.

For beginners and users who need simple map visualizations, Matplotlib is a good starting point. However, for more advanced mapping tasks, especially when working with geospatial data, Basemap or Cartopy are recommended. Cartopy, in particular, provides a more intuitive and modern API, making it easier to create complex maps and work with geospatial data.

Examples and Resources

To further explore mapping with Python libraries, here are a few examples and resources:

- **Example 1:** Plotting earthquakes on a world map using Cartopy and earthquake data from the US Geological Survey.
- **Example 2:** Creating a choropleth map of population density by state in the United States using Basemap and data from the U.S. Census Bureau.
- **Example 3:** Visualizing the distribution of COVID-19 cases across different countries using Matplotlib and data from public health organizations.

Here are some resources for further learning and reference:

- Matplotlib documentation: `https://matplotlib.org/`
- Basemap documentation: `https://matplotlib.org/basemap/`

- Cartopy documentation: `https://scitools.org.uk/cartopy/docs/latest/`
- EarthPy: A Python package for earth science data analysis and visualization, including mapping: `https://earthpy.readthedocs.io/en/latest/`
- GeoPandas: A Python library for working with geospatial data, including mapping: `https://geopandas.org/`

Conclusion

Python libraries provide powerful tools for mapping and visualizing geospatial data. In this section, we explored three popular libraries, Matplotlib, Basemap, and Cartopy, each offering different capabilities for creating maps. Matplotlib is a general-purpose plotting library suitable for basic map visualizations. Basemap provides more advanced functionalities specifically designed for mapping tasks. Cartopy builds upon Basemap and offers a modern and flexible interface for creating complex maps. Depending on the task at hand, one of these libraries can be chosen to create informative and visually appealing maps.

Advanced Data Visualization Techniques

In this section, we will explore advanced data visualization techniques that can be applied in geospatial data analysis. These techniques go beyond simple plotting and aim to provide more in-depth insights into spatial patterns and relationships. We will discuss various visualization methods and tools that can be used to effectively communicate complex geospatial information.

Multivariate Visualization

Multivariate visualization is a powerful technique that allows us to visualize patterns and relationships between multiple variables simultaneously. When working with geospatial data, it is often necessary to consider multiple attributes associated with each location. One way to address this is by using color coding or shading to represent different variables on a map. For example, we can use a color gradient to show the concentration of pollutant levels across different regions. This approach helps us identify spatial patterns and understand the correlations between variables.

To effectively represent multivariate data, we can use techniques such as parallel coordinates plots, scatter plots, and heatmaps. Parallel coordinates plots are especially useful when working with datasets that have a large number of variables. It allows us to visualize the relationships between variables by representing each data point as a line. Scatter plots can be used to visualize the relationships between two numeric variables, with each point representing a location on the map. Heatmaps provide a visual representation of the intensity or density of a particular attribute across different regions.

Interactive Visualization

Interactive visualization techniques enable users to dynamically explore and interact with geospatial data. Instead of presenting a static map, interactive visualizations allow users to manipulate parameters, zoom in and out, filter data, and view different layers or attributes on demand. This not only enhances the user experience but also enables users to gain deeper insights by visualizing data from different perspectives.

One popular tool for interactive geospatial visualization is D3.js (Data-Driven Documents). It is a JavaScript library that allows users to create dynamic and interactive visualizations directly in the browser. With D3.js, we can build interactive maps, scatter plots, bar charts, and other visualizations with smooth transitions and interactive elements. It also provides powerful data manipulation and transformation capabilities, making it a versatile tool for geospatial data analysis.

Another popular tool is the Leaflet library, which is specifically designed for interactive and mobile-friendly maps. Leaflet is lightweight, easy to use, and supports a wide range of spatial data formats. It allows users to add various layers, including tiled maps, markers, heatmaps, and overlay layers, to create interactive maps with ease. Leaflet also provides built-in support for user interaction, such as panning, zooming, and filtering data.

3D Visualization

While 2D maps are commonly used for geospatial data visualization, 3D visualization techniques can offer a more immersive and realistic representation of spatial data. 3D visualization allows us to visualize elevation, terrain, and other three-dimensional attributes, providing a more comprehensive view of the geographic features.

One popular library for 3D geospatial visualization is Three.js, a JavaScript library that makes it easy to create 3D scenes in the browser. With Three.js, we can visualize terrain, buildings, and other 3D objects on a globe or a custom map. It also provides support for lighting, shading, and animations, enabling us to create visually appealing and interactive 3D visualizations.

Another approach for 3D geospatial visualization is to utilize virtual reality (VR) or augmented reality (AR) technologies. These technologies offer a more immersive and interactive experience by allowing users to navigate and interact with the data in a virtual or augmented environment. VR and AR can be particularly useful for visualizing large-scale geospatial data, such as urban planning or land use simulations.

Geovisualization Tools and Libraries

In addition to the aforementioned tools, there are several other geovisualization libraries and tools available that can simplify the process of creating advanced visualizations. These tools provide a wide range of functionalities, including data processing, mapping, and interactive visualization.

Some popular geovisualization tools and libraries include:

- **Matplotlib:** A widely-used Python library for creating static, animated, and interactive visualizations. It provides a robust set of functions for creating maps, scatter plots, bar charts, and other types of visualizations.
- **geopandas:** An open-source Python library for manipulating geospatial data. It extends the popular pandas library with geospatial data types and operations, making it convenient for data preprocessing and analysis.
- **Folium:** A Python library for creating interactive maps using Leaflet.js. It allows users to create maps with various tile layers, markers, and shapefiles, and provides an easy-to-use interface for interactive exploration.
- **Carto:** A cloud-based geospatial platform that provides a range of tools for data visualization, analysis, and mapping. It offers an intuitive drag-and-drop interface and supports advanced features such as spatial SQL queries and geospatial analytics.
- **Tableau:** A popular data visualization tool that supports geospatial data visualization. It provides a user-friendly

interface for creating interactive dashboards and maps, and offers advanced data exploration and analysis capabilities.

These tools and libraries can greatly facilitate the process of creating advanced geospatial visualizations, regardless of the user's programming background or expertise. They provide a wide range of functionalities and allow users to create visually appealing and informative visualizations with minimum effort.

Example: Visualizing Crime Patterns in a City

To illustrate the application of advanced data visualization techniques, let's consider an example of visualizing crime patterns in a city. We have a dataset containing information about various crimes, including their types, locations, and timestamps. Our goal is to identify spatial and temporal patterns in crime occurrences and communicate them effectively.

We can start by creating a choropleth map that visualizes the overall crime rate in different regions of the city. We will use a color gradient to represent the crime rate, with darker shades indicating higher crime rates. This map will help us identify high-crime areas and understand the spatial distribution of crimes across the city.

Next, we can create a heatmap to visualize the temporal patterns in crime occurrences. The heatmap will show the density of crime incidents at different times of the day or week. This will help us identify peak crime hours or days and understand the temporal variations in crime rates.

To explore the relationship between crime types and locations, we can create a parallel coordinates plot. Each line on the plot corresponds to a crime incident, and the vertical axes represent different crime attributes (e.g., type, location, time). This plot will help us identify any correlations or patterns between crime types and their associated characteristics.

Finally, we can create an interactive visualization that allows users to explore the crime data dynamically. This can be done using a web-based tool like D3.js or Leaflet, where users can filter the data based on crime types, locations, or time periods, and visualize the filtered subset on a map.

By applying these advanced data visualization techniques, we can gain a comprehensive understanding of crime patterns in the city. These techniques will not only help us identify spatial and temporal clusters of crimes but also facilitate effective communication of the findings to stakeholders and policymakers.

Conclusion

Advanced data visualization techniques play a crucial role in geospatial data analysis by uncovering patterns, exploring relationships, and communicating insights effectively. In this section, we discussed multivariate visualization, interactive visualization, 3D visualization, and various geovisualization tools and libraries. We also provided an example of visualizing crime patterns in a city to demonstrate the practical application of these techniques.

By employing these techniques, analysts and researchers can gain deeper insights into geospatial data, detect hidden patterns, and make informed decisions based on the findings. Moreover, the availability of user-friendly tools and libraries makes it easier for individuals from diverse backgrounds to create compelling visualizations without extensive programming knowledge. As a result, advanced data visualization techniques continue to drive innovation and advancements in the field of geospatial data analysis.

Cartographic Design and Layout

Cartographic design and layout play a crucial role in presenting geospatial data effectively and accurately. In this section, we will explore the principles behind designing maps that are visually appealing, informative, and easy to interpret. We will also discuss the different elements of map layout and techniques for creating visually balanced and aesthetically pleasing maps.

Principles of Cartographic Design

Effective cartographic design requires an understanding of visual hierarchy, color theory, typography, and composition. By following these principles, maps can effectively convey information and engage the audience.

Visual Hierarchy

Visual hierarchy refers to the arrangement of map elements to guide the viewer's attention and understanding. It involves using visual cues such as size, color, contrast, and placement to prioritize important information. For example, larger and bolder features will naturally attract more attention than smaller and lighter ones. It is essential to consider the hierarchy of map elements when designing maps to ensure that the most critical information stands out.

Color Theory

Color plays a vital role in map design as it helps differentiate and represent different features on a map. Understanding color theory is essential to create visually appealing and meaningful maps. The appropriate use of color can enhance the readability and interpretation of the map. Some key principles to consider are:

- Use a limited color palette to avoid overwhelming the viewer.
- Choose colors that have contrast with each other and the background.
- Utilize colors in harmony to convey meaningful patterns or categories.
- Use color to represent qualitative or quantitative data effectively.

Color ramps and scales can also be used to represent gradients or ranges of data on a map, creating visual patterns that aid in interpretation.

Typography

Typography is the art of arranging type, including fonts, sizes, spacing, and hierarchy. Proper use of typography ensures that text on the map is easy to read and complements the overall design. Here are some best practices for typographic design:

- Use legible fonts and appropriate sizes for different map elements.
- Create typographic hierarchy by varying font sizes, weights, and styles.
- Align text elements consistently to maintain visual harmony.
- Avoid excessive use of decorative or intricate fonts that may hinder readability.

Typography should be used to enhance the clarity and organization of the map without overwhelming the viewer.

Composition

Composition refers to the arrangement and organization of all map elements within the map space. A well-composed map is visually balanced and creates harmony between the various components. Some considerations for map composition include:

- Create a focal point or main area of interest that draws the viewer's attention.
- Use visual elements such as lines, shapes, and patterns to guide the viewer's eye.
- Balance visual weight across the map to avoid clutter or unevenness.
- Follow the rule of thirds or other composition techniques to achieve a pleasing layout.

Careful composition ensures that the map conveys information in a clear and visually appealing manner.

Map Layout

Map layout refers to the arrangement of map elements on a page or a digital interface. An effective map layout should consider spatial relationships between different components, provide sufficient white space, and avoid clutter. Here are some important elements of map layout:

Legend and Scale

The map legend explains the symbols, colors, and other visual elements used in the map. It helps the viewer understand the meaning behind different map features. A well-designed legend should be clear, concise, and placed in a prominent location on the map to easily correlate with map elements.

The scale bar is another crucial element that represents the relationship between distances on the map and actual distances on the ground. It helps viewers understand the spatial extent and relative sizes of features. The scale bar should be appropriately sized and positioned for easy interpretation.

North Arrow and Compass Rose

The north arrow or compass rose indicates the orientation of the map, usually pointing towards the North or true North. It ensures that the viewer can interpret the map accurately. The north arrow should be placed in a consistent and visible location on the map layout.

Title and Annotations

A descriptive and informative title is essential for a map as it provides context and conveys the main theme or purpose of the map. The title should be clear, concise, and positioned prominently at the top or center of the map layout.

Annotations are text labels or callouts that provide additional information or explanations about specific map features. They can be used to highlight important points, provide context, or identify specific locations on the map. Annotations should be placed strategically to avoid clutter and ensure clarity.

Insets and Overviews

Insets are small maps that provide detailed views of specific areas or regions within a larger map. They are especially useful when dealing with maps of extensive areas or complex features. Insets should be placed near the related main map area and clearly labeled.

Overviews, on the other hand, are small-scale maps that show the entire extent or context of a larger map. They provide viewers with a quick reference and help them understand the spatial relationships of different regions. Overviews should be positioned strategically and may include features such as extent rectangles or arrows.

White Space and Alignment

White space refers to the empty or unmarked areas between map elements. It is crucial for creating visual breathing room, separating map portions, and avoiding clutter. Proper white space allows viewers to focus on essential map elements and improves overall readability.

Alignment is key to achieving a cohesive and organized map layout. Consistent alignment of elements along a grid or axis helps create visual harmony and structure. Aligning elements provides a sense of order and professionalism to the map design.

Designing Effective Maps

Designing effective maps requires a combination of technical skills, knowledge of geospatial data, and an understanding of cartographic design principles. Here are some tips and best practices to create visually appealing and informative maps:

- Understand the intended audience and purpose of the map, and design accordingly.

- Simplify and declutter the map by removing unnecessary elements and details.

- Make use of appropriate symbology and color schemes to represent data accurately.

- Ensure that all textual elements are legible and properly placed.
- Use proper map projection and scale to maintain accuracy and avoid distortions.
- Pay attention to spatial relationships and create visual hierarchies as per the importance of information.
- Regularly test and seek feedback on map designs to ensure effectiveness.

Remember that map design is both an art and a science. It requires creativity and thoughtful consideration of the data being represented. Strive to create maps that are not only visually appealing but also effectively communicate information to your audience.

Resources and Further Reading

- Brewer, Cynthia A. *Designing Better Maps: A Guide for GIS Users*. Esri Press, 2005.
- Dent, Borden D., Torguson, Jeffrey S., and Hodler, Thomas W. *Thematic Cartography and Geovisualization*. Prentice Hall, 2008.
- Krygier, John and Wood, Denis. *Making Maps: A Visual Guide to Map Design for GIS*. The Guilford Press, 2011.
- Robinson, Arthur H., Morrison, Joel L., and Muehrcke, Phillip C. *Elements of Cartography*. Wiley, 1995.

These resources provide further guidance on cartographic design principles, techniques, and best practices. They delve into the art and science behind creating effective maps and can serve as valuable references for refining your map design skills.

Exercises

1. Take a map of your choice and analyze the visual hierarchy used. Identify the different elements that contribute to the visual hierarchy and explain their significance.

2. Select a map and evaluate its color scheme. Assess how effectively the color palette represents the data on the map. Suggest possible improvements based on color theory principles.

3. Choose a small section of a map and experiment with different compositions. Create multiple variations by rearranging the elements and assess their visual balance and hierarchy.

4. Design a map layout that incorporates a legend, scale bar, title, and annotations. Pay attention to positioning and formatting to ensure clarity and readability.

5. Take a map that you have created or found and critically evaluate its effectiveness in conveying information. Consider the map's purpose, design principles, and user experience.

These exercises will help you apply the principles discussed in this section and enhance your skills in cartographic design and layout.

Conclusion

Designing effective maps involves careful consideration of visual hierarchy, color theory, typography, and composition. By adhering to these principles and incorporating map layout elements, you can create visually appealing and informative maps that effectively communicate geospatial data. Remember to experiment, seek feedback, and iterate on your designs to continuously improve your cartographic skills.

Spatial Regression and Machine Learning

Introduction to Spatial Regression

Spatial regression is a powerful technique used in geospatial data analysis to investigate and model relationships between spatially referenced variables. It allows us to understand how the variation in one variable is related to the variation in one or more other variables, taking into account the spatial component of the data.

In traditional regression analysis, we assume that the observations are independent of each other. However, in spatial regression, this assumption is violated due to the presence of spatial autocorrelation, which means that nearby locations tend to have similar values. Spatial regression models take into account this spatial autocorrelation and allow us to better capture the underlying spatial patterns and dependencies in the data.

Principles of Spatial Regression

Spatial regression models extend the traditional regression framework by incorporating spatial weights matrices into the model specification. These weight matrices define the spatial relationships between observations and quantify the influence of neighboring locations on each other. There are different ways to construct spatial weights matrices, such as contiguity-based weights (e.g., queen or rook contiguity) or distance-based weights (e.g., inverse distance or kernel weights).

The spatial autocorrelation in the dependent variable can be measured using spatial autocorrelation statistics such as Moran's I or Geary's C. These statistics provide a summary measure of how the values of a variable are spatially related. Positive spatial autocorrelation indicates that similar values are clustered together, while negative spatial autocorrelation suggests that dissimilar values are clustered.

In spatial regression analysis, we usually consider two main types of spatial dependence: spatial lag dependence and spatial error dependence. Spatial lag dependence occurs when the value of a variable at one location depends on the values of the same variable at neighboring locations. Spatial error dependence, on the other hand, captures the residual spatial dependence that is not explained by the spatial lag.

Model Specification

To incorporate spatial dependence into the regression model, we can introduce spatial lag terms or spatial error terms. The spatial lag model specifies that the dependent variable at a location is influenced by the values of the same variable at neighboring locations. The spatial error model, on the other hand, assumes that the residuals of the model have spatial autocorrelation.

The spatial lag model can be formulated as follows:

$$y = \rho W y + X\beta + \epsilon$$

where y is the dependent variable, X is the matrix of explanatory variables, β is the vector of regression coefficients, ϵ is the error term, and W is the spatial weights matrix. The coefficient ρ captures the spatial spillover effect, indicating the degree to which neighboring locations influence each other.

The spatial error model can be written as:

$$y = X\beta + \lambda WX + \epsilon$$

where λ represents the spatial autoregressive coefficient, capturing the effect of neighboring residuals on the dependent variable.

Estimation and Inference

To estimate the parameters of a spatial regression model, we can use maximum likelihood estimation or spatial two-stage least squares (S2SLS) estimation. Maximum likelihood estimation relies on the assumption of normally distributed errors, while the S2SLS estimator addresses potential endogeneity issues by instrumenting the endogenous variables using the lagged values or instruments derived from spatial eigenvector filtering.

In addition to parameter estimation, it is important to assess the goodness of fit of the spatial regression model. Common diagnostics include examining the residuals for spatial autocorrelation using Moran's I or Geary's C statistics. Residual plots and spatial correlograms can also be used to visualize the residual spatial patterns.

Applications and Examples

Spatial regression has diverse applications in various fields such as urban planning, environmental modeling, public health, and criminology. For example, in urban planning, spatial regression can be used to investigate the relationships between socioeconomic variables and patterns of urban growth. In environmental modeling, it can be employed to analyze the impact of land use on water quality or to model species distributions based on environmental variables.

Let's consider an example of using spatial regression to analyze the relationship between air pollution and socioeconomic factors in a city. We can collect data on air quality measurements at different locations across the city, as well as socioeconomic indicators such as income, education, and population density. By employing spatial regression, we can model the spatial dependence in air pollution and assess how it is influenced by the socioeconomic variables.

Challenges and Considerations

When working with spatial regression models, there are several challenges and considerations that should be taken into account. One important issue is the choice of spatial weights matrix, as different choices can lead to different model results. It is crucial to carefully consider the underlying spatial relationships and spatial processes in the data when constructing the weights matrix.

Another challenge is dealing with spatial heterogeneity, where the relationships between variables may vary across different spatial units. This can be addressed by incorporating spatially-varying coefficients or by using hierarchical spatial models.

Furthermore, spatial regression models often assume linearity and normality of the regression relationships and errors, which may not hold in all cases. Nonlinear spatial regression models and models based on alternative distributions may be used to address these limitations.

Conclusion

Spatial regression is a valuable technique for analyzing spatially referenced data and modeling spatial relationships. By taking into account the spatial autocorrelation and dependence in the data, spatial regression allows us to better understand and predict the spatial patterns and processes. It has numerous applications in various domains and can provide valuable insights into the factors driving spatial phenomena.

In the next sections, we will explore various advanced techniques in spatial regression, such as spatial econometric models and spatial machine learning models. We will also delve into practical examples and case studies to illustrate the application of spatial regression in real-world scenarios.

Further Reading

To delve deeper into the topic of spatial regression, the following resources are recommended:

- Anselin, L., & Bera, A. (1998). Spatial dependence in linear regression models with an introduction to spatial econometrics. *Spatial Analysis: Modelling in a GIS Environment*, 237-244.

- Bivand, R. S., Pebesma, E. J., & Gómez-Rubio, V. (2013). *Applied spatial data analysis with R*. Springer.

- Rey, S. J., & Anselin, L. (2010). *PySAL: A Python library of spatial analytical methods*. The Review of Regional Studies, 10(1), 245-268.

These resources provide comprehensive coverage of spatial regression techniques, their implementation in software packages, and their application in various fields.

Machine Learning in Geospatial Data Analysis

In recent years, machine learning has emerged as a powerful tool for analyzing and extracting insights from geospatial data. By leveraging the capabilities of machine learning algorithms, analysts can identify complex patterns, relationships, and trends within spatial datasets that may not be easily discernible using traditional statistical methods. This section provides an introduction to the application of machine learning techniques in geospatial data analysis, highlighting their benefits, challenges, and key considerations.

Background

Machine learning is a branch of artificial intelligence that focuses on the development of algorithms capable of automatically learning patterns and making predictions or decisions without explicit programming. It relies on the concept of training a model using a large amount of data, where the model learns to generalize patterns and make predictions on unseen data. In the context of geospatial data analysis, machine learning algorithms can be applied to tasks such as classification, regression, clustering, spatial interpolation, and anomaly detection.

Benefits of Machine Learning in Geospatial Data Analysis

There are several advantages to using machine learning in geospatial data analysis:

- **Improved Predictive Power:** Machine learning algorithms can capture complex relationships and patterns in geospatial data, enabling more accurate predictions and insights.

- **Efficient Data Processing:** Machine learning techniques can handle large volumes of geospatial data and automate repetitive tasks, saving time and effort for analysts.

- **Flexibility and Adaptability:** Machine learning models can be trained on a wide range of geospatial datasets and can adapt to different problem domains, making them versatile tools for various applications.

- **Discovery of Hidden Patterns:** Machine learning algorithms can uncover previously unknown patterns and relationships within geospatial data, leading to new insights and discoveries.

- **Integration with Other Geospatial Tools:** Machine learning techniques can be seamlessly integrated with other geospatial analysis tools, such as GIS software, to enhance data analysis capabilities.

Challenges and Considerations

While machine learning offers great potential for geospatial data analysis, there are several challenges and considerations that analysts need to be aware of:

Data Quality: The success of machine learning models heavily depends on the quality of input data. Geospatial datasets may contain errors, outliers, or missing values, which can adversely affect the performance of machine learning algorithms. It is essential to preprocess and clean the data before training the models.

Feature Selection and Engineering: Geospatial datasets often contain a large number of features, and selecting the most relevant ones is crucial for model performance. Additionally, feature engineering, which involves creating new features or transforming existing ones, may be required to better capture spatial patterns.

Spatial Autocorrelation: Geospatial data often exhibits spatial autocorrelation, where nearby locations are more likely to have similar attributes. This violates the assumption of independence in many machine learning algorithms. Techniques such as spatial weighting or incorporating spatial lag variables can be employed to address this issue.

Model Interpretability: Some machine learning algorithms, such as deep neural networks, are considered black-box models, meaning their decision-making process is hard to interpret. For certain applications, interpretability is crucial, and simpler, more transparent models may be preferred.

Data Imbalance: Geospatial datasets may suffer from class imbalance, where certain classes are underrepresented. This can lead to biased predictions and reduced model performance. Techniques like oversampling, undersampling, or the use of weighted loss functions can help mitigate this issue.

Computational Requirements: Training machine learning models on large geospatial datasets can be computationally intensive. Considerations need to be given to computational resources, such as memory requirements, parallel processing capabilities, and optimization techniques for efficient model training and evaluation.

Popular Machine Learning Techniques

A wide range of machine learning algorithms can be applied to geospatial data analysis. Some of the popular techniques include:

Random Forest: Random Forest is an ensemble machine learning algorithm that combines multiple decision trees to make predictions. It can handle high-dimensional data, capture nonlinear relationships, and provide feature importance estimates. Random Forest is commonly used for classification and regression tasks in geospatial analysis.

Support Vector Machines (SVM): SVM is a powerful supervised learning algorithm that can be used for both classification and regression tasks. SVM finds an optimal hyperplane that separates classes or predicts continuous values. SVMs have been applied to geospatial tasks such as land cover classification, urban growth prediction, and spatial interpolation.

Convolutional Neural Networks (CNN): CNNs are deep learning models primarily used for image analysis and recognition. They employ specialized layers to detect hierarchical spatial features. CNNs have been successfully applied to tasks such as land cover mapping, object detection, and change detection using remote sensing imagery.

K-means Clustering: K-means is an unsupervised learning algorithm used for clustering analysis. It aims to partition data into K clusters, where each observation belongs to the cluster with the nearest mean. K-means clustering has been employed for segmenting remotely sensed data, identifying land cover classes, and detecting spatial patterns.

Real-World Applications

Machine learning has found numerous applications in geospatial data analysis across various domains. Here are a few examples:

Urban Planning: Machine learning can assist in predicting urban growth, identifying suitable locations for infrastructure development, and optimizing transportation networks.

Environmental Monitoring: Machine learning algorithms can analyze satellite imagery to detect land cover changes, monitor deforestation, and assess the impact of climate change on ecosystems.

Public Health: Machine learning techniques can help analyze disease patterns, predict disease outbreaks, and optimize the allocation of healthcare resources in geographic areas.

Natural Hazards Management: Machine learning algorithms can be used to predict and mitigate the impact of natural disasters, such as hurricanes, earthquakes, and wildfires.

Conclusion

Machine learning has revolutionized geospatial data analysis by providing powerful tools for extracting insights, predicting patterns, and making informed decisions. By combining the strengths of machine learning algorithms with the rich spatial information available in geospatial datasets, analysts can unlock the full potential of these data sources. However, proper data preprocessing, feature engineering, spatial considerations, and model selection are critical for successful application. As technology advances and new methods emerge, the field of machine learning in geospatial analysis will continue to evolve, opening up new opportunities for understanding and managing our complex spatial world.

Resources and Further Reading

- Goodchild, M. F., Janelle, D. G. (Eds.) (2004). Spatially Integrated Social Science. Oxford University Press.

- Longley, P. A., Goodchild, M. F., Maguire, D. J., Rhind, D. W. (2015). Geographic Information Science & Systems. Wiley.

- Arribas-Bel, D., Openshaw, S. (2018). *Geocomputation: A Practical Primer*. SAGE Publications.

- Gatrell, A. C. (Ed.) (2013). *Geographies of Health: An Introduction* (3rd Edition). Wiley.

- Li, D., Dragićević, S. (Eds.) (2020). *Geospatial Data Science Techniques and Applications.* Springer.

Spatial Econometric Models

Spatial econometric models are statistical models that incorporate both spatial dependence and spatial heterogeneity into the analysis of spatially referenced data. These models are widely used in various fields such as economics, geography, urban planning, and environmental science to study the relationships between economic or social phenomena and their spatial context. In this section, we will explore the concepts, theory, and methods behind spatial econometric models.

Introduction to Spatial Econometrics

Traditional econometric models assume that observations are independent of each other, which may not be appropriate when dealing with spatial data. Spatial econometric models, on the other hand, acknowledge the existence of spatial dependence, which refers to the tendency of nearby locations to exhibit similar behaviors or outcomes. Spatial dependence can arise due to various mechanisms, such as spatial diffusion, spatial spillover effects, or the presence of common factors that affect neighboring locations.

To illustrate the concept of spatial dependence, consider a dataset of housing prices in different neighborhoods in a city. In traditional econometric models, the assumption of independence implies that the price of a house in one neighborhood is not affected by the prices of houses in neighboring neighborhoods. However, in reality, we often observe spatial autocorrelation, where houses in nearby neighborhoods tend to have similar prices. Spatial econometric models provide a framework to capture and quantify such spatial dependence.

Spatial Lag Models

One of the fundamental spatial econometric models is the Spatial Lag Model (SLM). The SLM assumes that the dependent variable is not only influenced by its own covariates but also by the values of the dependent variable in neighboring observations. Mathematically, the SLM can be represented as:

$$y = \rho W y + X\beta + u$$

where: - y is the dependent variable - ρ is the spatial autocorrelation parameter - W is the spatial weights matrix that captures the spatial relationship between observations - X is the matrix of covariates - β is the vector of coefficients associated with the covariates - u is the error term

The spatial lag term, $\rho W y$, captures the spatially dependent component of the dependent variable. The spatial weights matrix, W, defines the strength of the spatial relationship between observations. There are various ways to define the spatial weights matrix, such as using contiguity-based weights (e.g., queen contiguity or rook contiguity) or distance-based weights (e.g., inverse distance weighting or kernel weights).

Estimating the parameters of the SLM involves solving a system of linear equations, which can be done using maximum likelihood estimation or spatial two-stage least squares estimation. The coefficients of the spatial lag term (ρ) provide insights into the extent and direction of spatial dependence in the data. A positive ρ value indicates positive spatial autocorrelation, suggesting that similar values tend to cluster together, whereas a negative ρ value suggests negative spatial autocorrelation, indicating that similar values tend to repel each other.

Spatial Error Models

Another commonly used spatial econometric model is the Spatial Error Model (SEM). In the SEM, the spatial autocorrelation is explicitly modeled through the error term rather than the dependent variable itself. The SEM assumes that the error term is spatially autocorrelated, which means that the errors in nearby locations are correlated. Mathematically, the SEM can be represented as:

$$y = X\beta + \varepsilon$$

$$\varepsilon = \lambda W\varepsilon + u$$

where: - y, X, β, and u have the same meaning as in the SLM - ε is the error term, which consists of a spatially autocorrelated component and a non-spatially autocorrelated component - λ is the spatial autocorrelation parameter of the error term

The spatial error term, $\lambda W\varepsilon$, captures the spatial dependence in the error term. Estimating the parameters of the SEM involves solving a system of equations, which can be done using maximum likelihood estimation or instrumental variable estimation.

Compared to the SLM, the SEM allows for spatial heterogeneity in the relationship between the covariates and the dependent variable. The coefficients in the SEM represent the average effect of the covariates across all spatial units, while the spatial error term captures the spatially varying effects that are not explained by the covariates. The presence of spatial error autocorrelation can have implications on the estimation and inference of the coefficients in the SEM.

Spatial Durbin Models

The Spatial Durbin Model (SDM) extends the SLM and SEM by simultaneously incorporating both the spatial lag and spatial error components. The SDM allows for investigating the direct and indirect effects of the explanatory variables on the dependent variable. Mathematically, the SDM can be represented as:

$$y = \rho W y + X\beta + \delta W X\beta + \varepsilon$$

where: - y, X, β, and ε have the same meaning as in the SLM and SEM - δ is the spatial autocorrelation parameter of the spatially lagged covariates ($WX\beta$)

The presence of the spatially lagged covariates term, $\delta WX\beta$, allows for capturing the direct and indirect spatial effects of the explanatory variables on the dependent variable. Estimating the parameters of the SDM can be done using maximum likelihood estimation or generalized method of moments estimation.

The SDM provides a flexible framework for analyzing spatially dependent data by integrating both the spatial lag and spatial error components. By considering the direct and indirect spatial effects, the SDM offers a more comprehensive understanding of the relationships between variables in a spatial context.

Applications and Considerations

Spatial econometric models have a wide range of applications in various fields. For example, in urban economics, spatial econometric models can be used to study the spillover effects of housing prices or crime rates across neighborhoods. In environmental economics, these models can be applied to analyze the diffusion of pollution or the impact of environmental regulations on neighboring regions. In public health, spatial econometric models can help understand the spread of diseases or the accessibility to healthcare facilities.

When applying spatial econometric models, researchers need to consider several important considerations. First, the choice of spatial weights matrix is crucial, as it defines the spatial relationship between observations. The selection of an appropriate weights matrix should reflect the underlying spatial process and be driven by theory or empirical evidence. Second, the presence of spatial heterogeneity or non-stationarity should be carefully addressed, as the estimated coefficients may vary across spatial units or regions. Robust estimation methods or spatial panel data models can be employed to deal with such issues.

Lastly, it is important to interpret the results of spatial econometric models cautiously. Unlike traditional econometric models, spatial econometric models do not imply causality. The estimated coefficients represent associations rather than causal effects. Additional evidence, such as instrumental variable approaches or spatial randomization tests, may be required to establish causal relationships.

Conclusion

Spatial econometric models provide a powerful framework for analyzing and understanding spatially dependent data. They allow for the incorporation of spatial dependence and spatial heterogeneity into the analysis, offering valuable insights into the relationships between variables in a spatial context. By extending traditional econometric models, spatial econometric models enable researchers to investigate the direct and indirect effects of covariates on the dependent variable, considering the spatial autocorrelation in both the dependent variable and the error term. However, careful consideration of spatial weights, spatial heterogeneity, and interpretation of results is necessary when applying these models in practice.

Additional Resources

To further explore spatial econometric models, the following resources are recommended:

- Anselin, L. (1988). Spatial Econometrics: Methods and Models. Springer.
- LeSage, J. P., & Pace, R. K. (2009). Introduction to Spatial Econometrics. CRC Press.
- Arbia, G. (2014). A Primer for Spatial Econometrics: With Applications in R. Palgrave Macmillan.
- Rey, S. J., & LeSage, J. P. (2010). The Oxford Handbook of Urban Economics and Planning. Oxford University Press.

These resources provide comprehensive coverage of spatial econometric models, their estimation techniques, and applications in various fields. Additionally, software packages such as PySAL, GeoDa, and R libraries like spdep and splm offer tools and functions for implementing spatial econometric analysis.

Spatial Machine Learning Models

In recent years, there has been a growing interest in the intersection of machine learning and geospatial analysis. This has led to the development of spatial machine learning models, which aim to incorporate the spatial component of data into the learning process. These models are particularly useful in geospatial data analysis, where the spatial relationships between observations play a crucial role.

Background

Traditional machine learning models often assume that data points are independent and identically distributed (IID). However, in geospatial data analysis, this assumption is often violated due to spatial autocorrelation, where nearby observations tend to be more similar than distant ones. Spatial machine learning models address this issue by accounting for the spatial dependence in the data.

The goals of spatial machine learning models include predicting values at unobserved locations, identifying spatial patterns and relationships, and understanding the underlying processes that generate the spatial patterns.

Spatial Regression

One common approach to spatial machine learning is spatial regression. In spatial regression, the goal is to model the relationship between a dependent variable and a set of independent variables, while accounting for the spatial dependence. Spatial regression models often take the form:

$$y = \beta X + \epsilon \tag{1}$$

where y is the dependent variable, X is the matrix of independent variables, β is the vector of coefficients to be estimated, and ϵ is the error term.

To incorporate the spatial component, various spatial weights matrices can be used to capture the spatial relationships between the observations. One common choice is the spatial contiguity matrix, which defines the neighborhood structure based on spatial adjacency. The spatial weights matrix is then included in the regression model to account for the spatial autocorrelation. The model can be estimated using techniques such as maximum likelihood estimation or Bayesian methods.

Spatial Machine Learning Algorithms

In addition to spatial regression, there are several machine learning algorithms that have been adapted for spatial data analysis. These algorithms aim to capture the spatial patterns and dependencies in the data while making accurate predictions.

One popular algorithm is the kriging method, which originates from geostatistics. Kriging is a spatial interpolation technique that estimates values at unobserved locations based on the spatial relationships between the observed locations. It uses a weighted average of neighboring observations, where the weights are determined based on the spatial autocorrelation structure. Kriging can be used for both regression (ordinary kriging) and classification (kriging with a threshold).

Another spatial machine learning algorithm is the random forest. Random forest builds an ensemble of decision trees, where each tree is trained on a random subset of the data. In the spatial context, random forest can capture the spatial dependencies by including the spatial coordinates as additional predictor variables. This allows the algorithm to learn and exploit the spatial information to make accurate predictions.

Challenges and Considerations

Spatial machine learning models come with their own set of challenges and considerations. One key challenge is the choice of spatial weights matrix or neighborhood structure. Different weighting schemes can lead to different model results, and it is important to consider the underlying spatial relationships and patterns in the data when defining the weights.

Another consideration is model interpretability. While machine learning models are often praised for their predictive power, they can be difficult to interpret. This becomes even more challenging in the spatial context, where the relationships between variables are not only influenced by their own properties but also by their spatial location and the spatial relationships with other variables.

Additionally, the spatial component adds computational complexity to the modeling process. The large size of geospatial datasets and the need to perform calculations based on spatial coordinates can significantly increase the computational time and resource requirements.

Case Study: Predicting Housing Prices

To illustrate the concept of spatial machine learning models, let's consider a case study of predicting housing prices in a city. We have a dataset of housing sales with attributes such as location, number of bedrooms, square footage, and sale price. Our goal is to build a model that can accurately predict the sale price of houses at unobserved locations.

We can start by defining a spatial weights matrix based on the proximity between house locations. We can then use this matrix in a spatial regression model, where the dependent variable is the sale price and the independent variables are the attributes of the houses. By including the spatial weights matrix, we can account for the spatial autocorrelation and capture the spatial patterns in house prices.

Alternatively, we can use a spatial machine learning algorithm such as random forest. In this case, we would include the spatial coordinates of the houses as additional predictor variables, allowing the algorithm to learn the spatial patterns in house prices.

By building and evaluating different spatial machine learning models, we can identify the most accurate model for predicting housing prices in the city. This information can be valuable for real estate professionals, urban planners, and policymakers.

Further Reading

To delve deeper into spatial machine learning models, the following resources are recommended:

- *Spatial Regression Models* by Colin D. Wood
- *Introductory Econometrics for Spatial Analysis* by Luc Anselin
- *Machine Learning for Spatial Environmental Data: Theory, Applications, and Software* edited by Alexander Brenning
- *Geospatial Analysis: A Comprehensive Guide* by Michael J. de Smith, Michael F. Goodchild, and Paul A. Longley

These resources provide comprehensive coverage of the theory, methodologies, and applications of spatial machine learning models, along with practical examples and case studies.

Conclusion

Spatial machine learning models offer a powerful toolset for geospatial data analysis. By incorporating the spatial component into the learning process, these models can capture spatial patterns, relationships, and dependencies that traditional machine learning models may overlook. With the increasing availability of geospatial data and the advancement of machine learning techniques, spatial machine learning is becoming an essential tool for analyzing and understanding complex geospatial phenomena.

Geospatial Network Analysis

Introduction to Geospatial Networks

Geospatial networks, also known as spatial networks or network graphs, are a fundamental concept in geospatial data analysis. They are used to model and analyze the connectivity and relationships between geographic entities, such as cities, roads, rivers, and power grids. Geospatial networks enable us to understand and analyze complex spatial systems and phenomena, including transportation routes, communication networks, social interactions, and ecological networks.

In this section, we will explore the principles and techniques of geospatial network analysis. We will discuss the key concepts, methods, and tools used in this field, as well as real-world applications and challenges. By the end of this section, you will have a solid understanding of geospatial networks and be able to apply them to various geospatial analysis tasks.

Types of Geospatial Networks

Geospatial networks can be classified into several types based on the nature of the geographic entities and their relationships. The main types of geospatial networks include:

1. **Transportation Networks:** These networks represent the connectivity between locations primarily for transportation purposes. Examples include road networks, railway networks, and airline networks. Transportation networks are crucial for optimizing routes, analyzing traffic patterns, and planning urban transportation systems.

2. **Communication Networks:** These networks represent the flow of information between locations. Examples include telephone networks, internet networks, and social media networks. Communication networks help us understand information dissemination patterns, network congestion, and social connections.

3. **Ecological Networks:** These networks represent the ecological relationships between different natural entities, such as species, habitats, and food webs. Ecological networks are essential for understanding species distribution, biodiversity, and ecosystem dynamics.

4. **Utility Networks:** These networks represent the infrastructure systems responsible for providing essential services, such as water, electricity, and gas. Utility networks are critical for infrastructure planning, maintenance, and disaster management.

While these types of geospatial networks differ in their specific characteristics, they share common principles and analytical techniques.

Network Representation

In geospatial network analysis, the first step is to represent the network entities and their relationships. There are two primary ways to represent geospatial networks: node-based representation and edge-based representation.

1. **Node-based Representation:** In this representation, the network is represented using nodes (also known as vertices) and their relationships. Each node corresponds to a geographic entity, and the relationships between nodes are represented by connecting edges. This representation is useful when the focus is on the nodes and their attributes, such as analyzing social networks or transportation hubs.

2. **Edge-based Representation:** In this representation, the network is represented using edges and their connectivity. Each edge corresponds to a connection between two geographic entities, and the attributes of the entities are not explicitly represented. This representation is useful when the focus is on the connectivity and flow of entities, such as analyzing transportation routes or communication networks.

The choice of representation depends on the specific analysis objectives and the characteristics of the geospatial network under study.

Network Analysis Techniques

Geospatial network analysis involves a variety of techniques to analyze the connectivity, accessibility, and flow within and between network entities. Here are some key techniques used in geospatial network analysis:

1. **Network Measures:** These measures quantify various aspects of network structure and connectivity, such as degree centrality, betweenness centrality, and average shortest path length. Network measures help us

understand the importance and influence of different network entities and identify key hubs or bottlenecks.

2. **Route Analysis:** Route analysis focuses on finding optimal or efficient paths for moving between network entities. This includes shortest path analysis, which determines the shortest route between two entities, and network routing algorithms, which generate feasible paths considering various constraints (e.g., avoiding traffic or optimizing for different modes of transportation).

3. **Network Clustering:** Network clustering techniques identify groups or communities of closely connected entities within a network. Clustering helps reveal spatial patterns of similarity or dissimilarity between entities and is useful for understanding spatial segregation, functional regions, or social communities.

4. **Network Flow Analysis:** Network flow analysis examines the movement of resources, information, or entities through a network. This includes methods for modeling and optimizing flows, such as maximum flow algorithms and network equilibrium models. Network flow analysis is crucial for understanding congestion, capacity planning, and resource allocation.

These techniques provide powerful tools for analyzing and understanding the behavior and characteristics of geospatial networks.

Applications of Geospatial Network Analysis

Geospatial network analysis has wide-ranging applications in various domains. Here are a few examples:

- **Transportation Planning:** Geospatial network analysis helps optimize transportation routes, plan public transportation systems, and analyze traffic patterns. It enables efficient route planning for logistics, emergency response, and urban mobility.
- **Social Network Analysis:** Geospatial network analysis is applied to analyze social networks and understand social interactions. It helps identify influential individuals or groups, analyze information diffusion, and explore social segregation patterns.
- **Ecological Modeling:** Geospatial network analysis assists in understanding species distribution, habitat connectivity, and ecological corridors. It helps in designing conservation strategies, evaluating landscape connectivity, and assessing the impact of habitat fragmentation.
- **Infrastructure Management:** Geospatial network analysis is used for infrastructure management, including utility networks, telecommunications networks, and power grids. It aids in optimizing infrastructure layouts, analyzing network vulnerability, and planning for disaster management.

These applications demonstrate the broad significance of geospatial network analysis in solving real-world problems.

Challenges and Future Directions

While geospatial network analysis offers powerful analytical tools, there are several challenges to consider. Some of these challenges include:

- **Data Quality and Integration:** Geospatial networks require reliable and accurate spatial data. Integrating data from different sources and formats can be challenging due to inconsistencies and varying levels of data quality.

- **Scale and Complexity:** Geospatial networks can be vast and complex, requiring scalable and efficient algorithms and computational systems. Analyzing large networks with millions of nodes and edges poses computational and storage challenges.

- **Dynamic Networks:** Many geospatial networks are dynamic, changing over time due to various factors such as transportation demand, social interactions, or ecological processes. Analyzing dynamic networks introduces additional complexity and requires methods that capture temporal dynamics.

As geospatial technology advances, future directions in network analysis include:

- **Integration of Big Data:** With the proliferation of data from various sources, the integration of big data with geospatial networks will provide valuable insights and enable more accurate and comprehensive analysis.

- **Machine Learning and AI:** Applying machine learning and artificial intelligence techniques to geospatial network analysis will enhance prediction, network optimization, and decision-making capabilities.

- **Real-time Analysis:** Advancements in real-time data collection and analysis will enable the monitoring and management of geospatial networks in real-time, improving responsiveness and decision-making.

Conclusion

Geospatial networks are powerful tools for understanding and analyzing complex spatial relationships and systems. In this section, we have introduced the key concepts, methods, and applications of geospatial network analysis. We explored different types of geospatial networks, network representation techniques, analysis methods, and real-world applications.

By applying the principles and techniques of geospatial network analysis, you will be equipped to tackle a wide range of geospatial analysis tasks, from optimizing transportation routes to understanding social interactions and ecological systems. As technology advances and new challenges emerge, geospatial network analysis will continue to evolve as a key discipline in the field of geospatial data analysis and mapping with Python.

Resources for Further Reading:

- Mitchell, A. (2005). *The ESRI Guide to GIS Analysis, Volume 2: Spatial Measurements and Statistics.*
- Jiang, B. (2013). *Network Analysis in Geographic Information Systems.*
- Barthelemy, M. (2011). *Spatial Networks.*

Exercise:

1. Choose a real-world geospatial network, such as a transportation network or social network, and describe its key characteristics and potential applications.
2. Using Python and a suitable geospatial library, load a network dataset of your choice and visualize it using an appropriate network representation.

3. Perform basic network analysis tasks on the dataset, such as calculating degree centrality or finding the shortest path between two nodes.

Trick: When analyzing complex geospatial networks, simplifying the network representation by aggregating or generalizing nodes or edges can enhance computational efficiency without significantly affecting the analysis results.

Caveat: When using network analysis techniques, be cautious of potential biases and limitations in the data. The accuracy and comprehensiveness of the network data directly affect the validity and reliability of the analysis results.

Network Analysis in Python

Network analysis is a powerful technique for studying relationships and interactions between entities in a variety of fields, such as social sciences, transportation, biology, and computer science. In this section, we will explore how to perform network analysis using Python. We will cover the essential concepts, algorithms, and libraries to help you understand, manipulate, and analyze network data effectively.

Introduction to Network Analysis

A network, also known as a graph, consists of a set of nodes (also called vertices) and a set of edges (also called links or connections) that connect pairs of nodes. Networks can be represented mathematically as graphs, where nodes are represented by points or circles, and edges are represented by lines connecting the nodes.

Network analysis involves studying the structure, properties, and behavior of networks. It allows us to analyze complex systems by examining the relationships between components and identifying patterns and properties that emerge from these relationships.

In Python, there are several powerful libraries available for network analysis, such as NetworkX, igraph, and graph-tool. In this section, we will focus on NetworkX, a popular Python library for studying the structure and dynamics of networks.

NetworkX: A Python Library for Network Analysis

NetworkX is a flexible and efficient library for network analysis in Python. It provides an extensive collection of tools for studying the structure and properties of networks, as well as algorithms for network analysis and visualization.

To get started with NetworkX, you first need to install the library. You can install it using pip, the Python package installer, by running the following command in your terminal:

```
pip install networkx\index{networkx}\index{networkx}
```

Once you have installed NetworkX, you can import it into your Python program using the following line of code:

```
import\index{import} networkx\index{networkx} as nx
```

Creating a Network

Let's start by creating a basic network using NetworkX. We can create an empty network by creating an instance of the `Graph` class:

```
G = nx.Graph()
```

We can then add nodes to the network using the `add_node` method and edges using the `add_edge` method. For example, let's create a network with three nodes connected by two edges:

```
G.add_node(1)
G.add_node(2)
G.add_node(3)
G.add_edge(1, 2)
G.add_edge(2, 3)
```

Network Visualization

Visualizing networks can be helpful for understanding their structure and properties. NetworkX provides various methods for visualizing networks using different layouts, node colors, and edge styles.

To visualize a network using NetworkX, you can use the `draw` function. For example, let's visualize the network we created earlier:

```
import\index{import} matplotlib.pyplot as plt

nx.draw(G, with_labels=True, node_color='lightblue'

plt.show()
```

This code will display a graphical representation of the network, where nodes are labeled with their IDs, node color is set to light blue, and edge color is set to gray.

Network Analysis Algorithms

NetworkX provides a wide range of algorithms for analyzing networks. These algorithms can help us understand various network properties, such as centrality, connectivity, clustering, and community structure.

Let's explore some of the commonly used network analysis algorithms in NetworkX:

Centrality Measures

Centrality measures quantify the importance or influence of a node in a network. Some common centrality measures include degree centrality, betweenness centrality, and eigenvector centrality.

Degree Centrality Degree centrality measures the number of connections (edges) that a node has. It can be calculated using the `degree_centrality` function in NetworkX:

```
degree_centrality = nx.degree_centrality(G)
```

Betweenness Centrality Betweenness centrality measures the extent to which a node lies on the shortest paths between other nodes in the network. It can be calculated using the `betweenness_centrality` function:

```
betweenness_centrality = nx.betweenness_centrality(G)
```

Eigenvector Centrality Eigenvector centrality measures the importance of a node in a network based on the centrality of its neighbors. It can be calculated using the `eigenvector_centrality` function:

```
eigenvector_centrality = nx.eigenvector_centrality(G)
```

Connectivity

Connectivity measures the degree to which a network is connected or how easily information can flow between nodes. NetworkX provides various algorithms for computing connectivity measures, such as the number of connected components, the size of the largest connected component, and the average shortest path length.

Connected Components Connected components represent groups of nodes that are connected to each other but not connected to nodes in other groups. The `connected_components` function in NetworkX can be used to compute the connected components of a network:

```
connected_components = nx.connected_components(G)
```

Shortest Paths The shortest path between two nodes is the smallest number of edges that must be traversed to go from one node to the other. NetworkX provides the `shortest_path_length` function to compute the shortest path length between nodes:

```
shortest_path_length = nx.shortest_path_length(G, s
```

Clustering

Clustering measures the tendency of nodes in a network to form clusters or communities. NetworkX provides various algorithms for computing clustering coefficients, such as the average clustering coefficient, global clustering coefficient, and local clustering coefficient.

Clustering Coefficient The clustering coefficient of a node measures the degree to which its neighbors are connected to each other. The `clustering` function in NetworkX can be used to compute the clustering coefficient of each node in a network:

```
clustering_coefficient = nx.clustering(G)
```

Summary

In this section, we explored network analysis using Python and the NetworkX library. We learned about the basics of network analysis, the NetworkX library, creating networks, visualizing networks, and various network analysis algorithms.

Network analysis is a powerful tool for understanding the structure, behavior, and properties of networks in various domains. By applying network analysis techniques and algorithms, we can gain insights into complex systems and make informed decisions.

In the next section, we will delve into advanced topics in geospatial networks, where we will explore additional techniques and algorithms specifically tailored for geospatial data analysis. Stay tuned!

Exercises

1. Create a network using NetworkX with 10 nodes and 15 edges. Visualize the network using a suitable layout and node colors.

2. Compute the degree centrality and betweenness centrality of the nodes in the network created in exercise 1. Identify the most central nodes in terms of these centrality measures.

3. Compute the connected components of the network created in exercise 1. How many connected components are there? Identify the largest connected component.

4. Compute the average clustering coefficient of the network created in exercise 1. Interpret the result.

5. Find the shortest path between two randomly selected nodes in the network created in exercise 1. Calculate the shortest path length and visualize the path.

Resources and Further Reading

1. NetworkX documentation: `https://networkx.org/documentation` 2. Python for Data Analysis, 2nd Edition by Wes McKinney 3. Complex Network Analysis in Python: Recognize, Construct, Visualize, Analyze, and Interpret Complex Networks by Dmitry Zinoviev 4. Social Network Analysis for Startups: Finding Connections on the Social Web by Maksim Tsvetovat and Alexander Kouznetsov

Spatial Network Analysis Techniques

In this section, we will explore various techniques for spatial network analysis. Spatial networks are structures that represent the relationships between objects or locations within a geographic space. They are widely used in various domains, such as transportation, telecommunications, social networks, and urban planning. Spatial network analysis allows us to gain insights into connectivity patterns, flow dynamics, accessibility, and optimal routing, among other important factors.

Basic Concepts of Spatial Networks

Before diving into spatial network analysis techniques, let's first understand the basic concepts of spatial networks. A spatial network consists of two main components: nodes and edges. Nodes represent the locations or objects of interest, while edges represent the connections or relationships between nodes.

In spatial networks, nodes may correspond to cities, buildings, road intersections, or any other point of interest. Edges, on the other hand, represent the physical or logical connections between these nodes. For example, in a transportation network, edges could represent roads, railways, or air routes.

Spatial networks can be classified based on their characteristics and properties. Some common types of spatial networks include:

- Road networks: Representing the physical connections between different road segments and intersections.
- Social networks: Representing the relationships between individuals or groups, such as friendship networks or collaboration networks.
- Telecommunications networks: Representing the connections between cell towers, telephone lines, or internet routers.

Now that we have a basic understanding of spatial networks, let's explore some techniques for analyzing them.

Network Connectivity Analysis

One of the fundamental aspects of spatial network analysis is network connectivity analysis. It focuses on understanding the connectivity patterns within a spatial network and identifying the components or subnetworks that are connected or disconnected.

There are several measures that can be used to assess network connectivity. One common measure is the degree centrality, which quantifies the number of edges connected to a node. Nodes with high degree centrality are considered important in terms of network connectivity.

Another measure is the betweenness centrality, which quantifies the extent to which a node lies on the shortest paths between other nodes. Nodes with high betweenness centrality are often critical in terms of facilitating flow or communication between different parts of the network.

To identify connected or disconnected subnetworks, we can use graph traversal algorithms, such as breadth-first search or depth-first search. These algorithms allow us to explore the network starting from a given node and visit all the connected nodes.

Network Flow Analysis

Network flow analysis focuses on understanding the dynamics of flow within a spatial network. It involves studying how resources, information, or individuals move through the network and identifying the bottlenecks, congestion points, or optimal routes.

One common technique for network flow analysis is the shortest path analysis. It aims to find the shortest path between two nodes in a network based on certain criteria, such as distance, travel time, or cost. Dijkstra's algorithm is a popular algorithm for finding the shortest paths in a network.

Another important aspect of network flow analysis is the capacity of edges. Edges in a spatial network often have limited capacity, which represents the maximum flow that can pass through them. Analyzing the flow and capacity of edges allows us to identify potential congestion points or areas where the network needs to be expanded.

Network Clustering Analysis

Network clustering analysis focuses on identifying groups or clusters of nodes that are densely connected within themselves but sparsely connected with nodes in other clusters. It helps in understanding the underlying structure and organization of a spatial network.

There are various algorithms for network clustering analysis, such as Girvan-Newman algorithm or Louvain algorithm. These algorithms iteratively divide the network into clusters based on certain criteria, such as the betweenness centrality of edges. The resulting clusters provide insights into the modular structure of the network.

Network clustering analysis can be applied to various domains. For example, in social networks, it can help identify communities or groups of individuals with strong connections. In transportation networks, it can help identify clusters of road segments that are often used together.

Spatial Network Visualization

Visualization plays a crucial role in spatial network analysis as it allows us to effectively communicate the complex patterns and relationships within a network. Various visualization techniques can be used to represent spatial networks, such as node-link diagrams, matrix-based representations, or geographic maps.

Node-link diagrams are commonly used to represent spatial networks, where nodes are represented by points, and edges are represented by lines. The positions of nodes can be determined based on their geographical coordinates or through force-directed layout algorithms.

Matrix-based representations, such as adjacency matrices or distance matrices, can also be used to visualize the connections between nodes in a network. These matrices provide a compact representation of the connectivity patterns and can be useful for large networks.

Geographic maps can provide a visual representation of spatial networks by overlaying the network topology onto a map. This allows us to understand the spatial distribution of nodes and edges, and how they relate to the geographical context.

Real-World Applications

Spatial network analysis techniques find applications in various real-world scenarios. Here are a few examples:

- **Transportation Planning:** Spatial network analysis helps in designing efficient transportation systems, optimizing routes, and identifying congestion points.
- **Communications:** Network analysis techniques are used to optimize telecommunication networks, improve signal strength, and ensure efficient data transmission.
- **Social Networks:** Spatial network analysis is useful in understanding the dynamics of social networks, identifying influential individuals or groups, and predicting trends.
- **Epidemiology:** Network analysis helps in tracking the spread of diseases, identifying critical nodes for intervention, and evaluating the impact of interventions.

Conclusion

Spatial network analysis techniques provide valuable insights into the connectivity patterns, flow dynamics, and structure of spatial networks. By understanding these techniques, researchers and practitioners can gain a deep understanding of complex systems and make informed decisions in various domains.

In this section, we explored the basic concepts of spatial networks, including nodes, edges, and different types of networks. We also discussed network connectivity analysis, flow analysis, clustering analysis, visualization techniques, and real-world applications. With this knowledge, readers will be well-equipped to analyze and visualize spatial networks effectively.

Advanced Topics in Geospatial Networks

In this section, we will explore advanced topics in geospatial networks, building upon the foundational concepts introduced in the previous sections. Geospatial networks are essential tools for analyzing and understanding spatial relationships and connectivity in various domains, including transportation systems, social networks, and infrastructure planning. We will delve into sophisticated techniques and methodologies that offer deeper insights into the structure, behavior, and dynamics of geospatial networks.

Spatial Network Resilience

Understanding the resilience of geospatial networks is crucial for planning and managing various systems, such as transportation networks or power grids. Resilience measures the ability of a network to withstand disruption or recover from it. In this context, disruptions can range from natural disasters to intentional attacks.

To analyze the resilience of geospatial networks, we need to consider various factors, including network topology, connectivity, and redundancy. One common measure of network resilience is the concept of network centrality, which quantifies the importance of nodes or edges within a network. Centrality metrics, such as degree centrality, betweenness centrality, or closeness centrality, can help identify critical nodes or edges that, if disrupted, would significantly impact network performance.

Another important aspect of network resilience is the concept of robustness. Robustness measures the ability of a network to maintain its functionality even when certain nodes or edges are removed. By simulating node or edge failures and evaluating the network's performance, we can identify critical components and design strategies to enhance resilience.

Spatial Network Dynamics

Geospatial networks are not static; their structure and behavior evolve over time. Analyzing the temporal dynamics of geospatial networks allows us to better understand their growth patterns, identify emerging patterns, and predict future behavior. Here, we will discuss two key aspects of spatial network dynamics: network evolution and network diffusion.

Network evolution refers to the changes in network structure over time. Understanding how geospatial networks grow and evolve can help predict future connectivity patterns and guide infrastructure planning. This can be achieved through various approaches, such as preferential attachment models, where new nodes are more likely to connect to well-connected nodes, or network growth models based on geographical proximity.

Network diffusion focuses on how information, objects, or phenomena spread across a network. Diffusion processes are prevalent in various domains, such as the spread of diseases, information dissemination, or product adoption. Analyzing network diffusion allows us to identify influential nodes for spreading or blocking information and to model the dynamics of contagious processes.

Spatial Network Optimization

Optimizing geospatial networks involves finding the best configuration or allocation of network components to achieve specific objectives. In this context, optimization can refer to different goals, such as improving accessibility, minimizing travel time, or maximizing network robustness.

One common optimization problem in geospatial networks is the *minimum spanning tree* (MST) problem, which aims to find the subset of edges that connects all nodes with the minimum total cost. MSTs are useful for designing efficient transportation or utility networks, ensuring connectivity with minimal investment.

Another important optimization problem is the *vehicle routing problem* (VRP), which deals with finding the most efficient routes for a fleet of vehicles to serve a set of locations. VRP has numerous applications in logistics, transportation planning, and distribution systems. Solving VRP involves considering factors such as vehicle capacity, time windows, and customer preferences, to minimize travel distance or time while satisfying constraints.

Complex Spatial Networks

In real-world scenarios, geospatial networks often exhibit complex characteristics that cannot be fully captured by simple network models. Complex spatial networks are characterized by non-linear relationships, hierarchical structures, and emergent properties. Understanding and analyzing complex spatial networks requires advanced computational tools and methodologies.

One such approach is *percolation theory*, which studies the critical properties and connectivity of complex networks. By modeling node or edge failures under different scenarios, percolation theory enables us to evaluate the robustness and fragility of geospatial networks.

Another technique used to analyze complex spatial networks is *multilayer network analysis*. Multilayer networks capture the interdependencies between different types of connections in a network, allowing for a more comprehensive understanding of complex systems. This approach has applications in various domains, such as transportation planning, urban studies, and social network analysis, where different types of relationships coexist.

Advanced Network Visualization

Visualizing geospatial networks is essential for effectively communicating complex information and insights derived from network analysis. Advanced network visualization techniques can help reveal patterns, highlight key features, and convey relationships within a network.

One powerful visualization approach is *graph layout.* Graph layout algorithms position nodes and edges in a meaningful way to emphasize the underlying structure and assist with pattern recognition. Various layout algorithms, such as spring embedding, force-directed placement, or spectral layout, can be employed based on the nature of the network and the desired visualization goals.

Another technique used in advanced network visualization is *interactive visualization.* Interactive visualizations enable users to explore and interact with the network data, revealing hidden patterns and facilitating data-driven decision making. Interactive visualization tools allow users to filter, sort, and query network data, providing a more nuanced and personalized exploration experience.

Conclusion

In this section, we have explored advanced topics in geospatial networks, expanding on the foundational concepts introduced earlier. We have discussed the resilience and dynamics of geospatial networks, optimization techniques, complex network analysis, and advanced visualization approaches. By understanding and applying these advanced topics, practitioners can gain deeper insights into the behavior and operation of geospatial networks, enabling them to make informed decisions and solve complex problems in various domains.

To further explore these advanced topics, we recommend the following resources:

- *Network Science* by Albert-László Barabási
- *Introduction to the Theory of Complex Systems* by Stefan Thurner
- *Geospatial Analysis: A Comprehensive Guide* by Michael J. de Smith, Michael F. Goodchild, and Paul A. Longley
- *Complex Networks: Structure, Robustness, and Function* by Ernesto Estrada and Philip A. Knight

We hope that the knowledge gained from this section empowers you to tackle real-world challenges using advanced techniques in geospatial network analysis.

Geospatial Data Analysis in the Cloud

Introduction to Cloud Computing and GIS

Cloud computing and geographic information systems (GIS) are two rapidly evolving technologies that have had a significant impact on the field of geospatial data analysis. In this section, we will delve into the fundamentals of cloud computing and its applications in GIS. We will explore the benefits of cloud computing for geospatial analysis, the different cloud service providers, and the tools and techniques available for geospatial analysis in the cloud.

Background of Cloud Computing

Cloud computing is the delivery of computing services over the internet on a pay-as-you-go basis. Instead of owning physical hardware and software infrastructure, cloud computing allows users to access and use computing resources provided by third-party service providers. These resources include servers, databases, storage, networking, and software applications. Cloud computing offers several advantages, such as scalability, cost-efficiency, flexibility, and reliability.

In the context of geospatial analysis, cloud computing provides a platform for storing, processing, and analyzing large volumes of geospatial data. It eliminates the need for users to invest in expensive hardware and software, which can be a significant barrier to entry for many organizations. Cloud computing also enables seamless collaboration and sharing of geospatial data and analysis results among multiple users across different locations.

Cloud Service Providers for Geospatial Analysis

There are several cloud service providers that offer infrastructure and services for geospatial analysis. The most prominent ones include Amazon Web Services (AWS), Google Cloud Platform (GCP), and Microsoft Azure. These cloud service providers offer a wide range of services and tools that are specifically designed for geospatial analysis.

AWS provides a comprehensive suite of services, such as Amazon S3 for storage, Amazon EC2 for virtual servers, and Amazon Redshift for data warehousing. It also offers specialized geospatial services, including Amazon Athena for querying geospatial data, Amazon EMR for big data processing, and Amazon GroundTruth for satellite imagery analysis.

GCP offers a similar set of services, such as Google Cloud Storage for storage, Google Compute Engine for virtual machines, and Google BigQuery for data analytics. It also provides geospatial-specific services, such as Google Earth Engine for analyzing satellite imagery and Google Maps Platform for location-based services.

Microsoft Azure provides services such as Azure Blob Storage for storage, Azure Virtual Machines for virtual servers, and Azure SQL Database for data management. It also offers geospatial services like Azure Maps for geocoding and routing, and Azure IoT Hub for managing geospatial data from IoT devices.

Geospatial Analysis in the Cloud

Geospatial analysis involves the processing, modeling, and visualization of geospatial data to gain insights and make informed decisions. Cloud computing provides an ideal platform for geospatial analysis due to its scalability and computing power. Here are some of the key activities involved in geospatial analysis in the cloud:

- **Data storage and management:** Cloud storage services enable users to store and manage large volumes of geospatial data. Users can choose from different storage options, such as object storage or relational databases, based on their specific requirements. Cloud storage also provides features like data backup, encryption, and access control.

- **Data preprocessing:** Geospatial data often requires preprocessing before analysis. Cloud computing offers powerful tools and services for data preprocessing, such as spatial data transformations, filtering, and aggregation. These tools enable users to clean and prepare geospatial data for further analysis.

- **Spatial analysis:** Cloud-based geospatial analysis platforms provide a wide range of spatial analysis capabilities. Users can perform various spatial operations, such as buffering, intersection, and spatial queries, to extract meaningful insights from geospatial data. Cloud-based platforms also offer built-in algorithms and models for spatial statistics, network analysis, and image analysis.

- **Data visualization:** Cloud computing provides tools for the visualization of geospatial data and analysis results. Users can create interactive maps, charts, and dashboards to communicate geospatial insights effectively. Cloud-based visualization tools offer features like layer styling, thematic mapping, and real-time updates.

- **Collaboration and sharing:** Cloud computing enables seamless collaboration and sharing of geospatial data and analysis results among multiple users. Users can work on the same project simultaneously, share their work with others, and collaborate on data analysis workflows. Cloud-based platforms also offer version control, permission management, and access logs to ensure data security and compliance.

- **Scalability and performance:** Cloud computing provides the scalability and computing power required for large-scale geospatial analysis. Users can dynamically scale up or down their computing resources based on the size and complexity of their analysis tasks. Cloud-based platforms also offer parallel processing capabilities, distributed computing frameworks, and GPU acceleration for high-performance geospatial analysis.

Challenges and Considerations

While cloud computing offers numerous benefits for geospatial analysis, there are some challenges and considerations that users need to be aware of:

- **Data security and privacy:** Storing geospatial data in the cloud raises concerns about data security and privacy. Users need to ensure that their data is encrypted, access-controlled, and protected from unauthorized access or data breaches. Compliance with data protection regulations, such as GDPR, is also critical.

- **Cost management:** Cloud computing services are typically charged on a usage basis. Users need to carefully plan their resource utilization to optimize costs. This includes selecting the right storage options, managing compute instances efficiently, and leveraging cost-saving measures like reserved instances or spot instances.

- **Data transfer and latency:** Uploading and downloading large volumes of geospatial data to and from the cloud can be time-consuming, especially over low-bandwidth networks. Users need to consider data transfer costs and latency when designing their geospatial analysis workflows.

- **Vendor lock-in:** Moving geospatial data and analysis workflows to the cloud may lead to vendor lock-in, where users become dependent on a specific cloud service provider. Users need to carefully evaluate the portability and interoperability of their geospatial data and analysis tools across different cloud platforms.

Conclusion

Cloud computing has revolutionized the field of geospatial analysis by providing scalable, cost-effective, and collaborative platforms for storing, processing, and analyzing geospatial data. With the availability of specialized geospatial services and tools provided by cloud service providers like AWS, GCP, and Azure, geospatial analysis has become more accessible and efficient. However, it is essential to consider the challenges and considerations associated with cloud computing, such as data security, cost management, and vendor lock-in. As cloud computing continues to evolve, it will play an increasingly central role in geospatial analysis and enable new possibilities for solving complex geospatial problems.

Resources and Further Reading

To learn more about cloud computing and its applications in geospatial analysis, you may refer to the following resources:

- Butler, D. L. (2019). Cloud-Based Services for Geospatial Data Analysis and Management. In Cloud Computing for Geospatial Big Data Analytics (pp. 19-39). Springer.
- Hsu, L., and Huang, X. (2017). Cloud-based GIS infrastructure for spatial big data analytics. International Journal of Geo-Information, 6(2), 45.
- Li, F., and Zhou, M. (2020). Cloud Computing in Geospatial Data Processing: Technologies and Applications. In Geospatial Computing in Mobile Devices (pp. 79-93). Springer.
- Mitchell, T., and Harden, E. (2019). Introduction to Cloud-Based Geospatial Data Analysis. In The GIS Guide to Public Domain Data (pp. 227-250). ESRI Press.

Amazon Web Services (AWS) for Geospatial Analysis

In recent years, the rise of cloud computing has revolutionized the way data is stored, processed, and analyzed. Amazon Web Services (AWS) is one of the leading cloud computing platforms, offering a wide range of services that can be utilized for geospatial analysis. In this section, we will explore some of the key services provided by AWS and how they can be leveraged for geospatial data analysis.

Introduction to AWS

Amazon Web Services (AWS) is a comprehensive cloud computing platform provided by Amazon. It offers a broad spectrum of services including compute power, storage, databases, analytics, machine learning, and more. AWS provides a flexible, scalable, and cost-effective solution for processing and analyzing large volumes of geospatial data.

Storage Services

One of the fundamental aspects of geospatial analysis is data storage. AWS offers various storage services that cater to different data requirements. Two key services are:

Simple Storage Service (S3)

Amazon S3 is an object storage service that allows you to store and retrieve large amounts of unstructured data. It provides durability, availability, and scalability, making it an ideal choice for storing geospatial datasets. S3 provides a simple RESTful API for data access and is compatible with a wide range of data formats.

Elastic File System (EFS)

Elastic File System (EFS) is a scalable and fully managed file storage service. It provides a shared file system that can be accessed concurrently by multiple compute instances. EFS is particularly useful when working with distributed geospatial analysis workflows, allowing multiple users to access and modify data simultaneously.

Compute Services

AWS offers a variety of compute services that enable the processing and analysis of geospatial data. The following services are particularly relevant:

Elastic Compute Cloud (EC2)

Amazon EC2 provides resizable compute capacity in the cloud. It allows you to create and configure virtual servers, known as instances, which can be customized to meet specific requirements. EC2 instances can be utilized for running geospatial analysis algorithms, such as spatial statistics or machine learning models.

Lambda

AWS Lambda is a serverless computing service that allows you to run code without provisioning or managing servers. It executes your code in response to events, such as data changes or user requests. Lambda functions can be used for on-demand geospatial data processing, enabling cost-effective and scalable analysis workflows.

Database Services

Geospatial analysis often requires efficient storage and retrieval of spatial data. AWS provides several database services that can handle geospatial datasets effectively:

Relational Database Service (RDS)

Amazon RDS is a managed relational database service that supports popular database engines such as PostgreSQL, MySQL, and SQL Server. These databases can be extended with geospatial data types and used for storing and querying spatial data. RDS offers high scalability, automated backups, and built-in monitoring for efficient management of geospatial databases.

Aurora

Amazon Aurora is a highly scalable and fully managed relational database service compatible with MySQL and PostgreSQL. It provides better performance and lower costs compared to traditional databases, making it suitable for geospatial applications that require high throughput and low latency.

Analytics and Machine Learning Services

Geospatial data analysis often involves performing complex analytical tasks and applying machine learning algorithms. AWS offers the following services to facilitate these tasks:

Elastic MapReduce (EMR)

Amazon EMR is a cloud-based big data platform that enables processing of large datasets using popular frameworks such as Apache Spark and Hadoop. EMR provides pre-configured clusters and optimized storage options for geospatial analysis. With EMR, you can perform distributed processing of spatial data, enabling efficient analysis of large-scale geospatial datasets.

SageMaker

Amazon SageMaker is a fully managed machine learning service that provides end-to-end functionality for building, training, and deploying machine learning models. It supports popular deep learning frameworks like TensorFlow and PyTorch, which can be applied to geospatial data for tasks such as image classification or object detection.

Integration with Geospatial Tools and Libraries

AWS services are designed to integrate smoothly with popular geospatial tools and libraries, allowing seamless integration into existing workflows. Some commonly used tools include:

GeoPandas

GeoPandas is an open-source Python library that extends the capabilities of Pandas dataframes to handle geospatial data. It provides geometric operations and spatial analysis capabilities. GeoPandas can be used in conjunction with AWS services for data transformation and analysis tasks.

GDAL

GDAL (Geospatial Data Abstraction Library) is a widely used open-source library for reading and writing raster and vector geospatial data formats. It offers a rich set of functionalities for data manipulation and transformation. GDAL can be used alongside AWS services to import or export geospatial datasets.

Cost Optimization and Scalability

One of the major advantages of AWS is its cost optimization and scalability features. By utilizing services such as Auto Scaling and Spot Instances, you can optimize costs by scaling compute resources based on demand. Additionally, AWS provides cost monitoring and management tools to help track and control geospatial analysis costs effectively.

Case Study: Real-Time Flood Monitoring

To illustrate the capabilities of AWS for geospatial analysis, let's consider a real-world case study of real-time flood monitoring. The goal is to analyze rainfall data in real-time and identify areas at risk of flooding.

Using AWS services, we can set up an EC2 instance to collect and process rainfall data from various sources. The data can be stored in Amazon S3 or a database service like Amazon RDS. By leveraging the Elastic MapReduce (EMR) service, we can perform distributed processing and extract insights from the data. Machine learning algorithms, implemented using Amazon SageMaker, can be used to predict and identify areas at risk of flooding based on historical data.

With the scalability and cost optimization features offered by AWS, it becomes feasible to analyze large volumes of real-time data efficiently. The results can be visualized using Python libraries like Matplotlib or web-based mapping tools like Leaflet, providing critical information to relevant stakeholders.

Conclusion

Amazon Web Services offers a wide range of services that can be leveraged for geospatial analysis. From storage and compute services to analytics and machine learning tools, AWS provides a flexible and scalable platform for processing and analyzing geospatial data. By integrating AWS services with geospatial tools and libraries, it becomes easier to perform complex analysis tasks and derive meaningful insights. Whether it's environmental monitoring, urban planning, or disaster management, AWS can be a valuable resource for geospatial data analysis.

Google Cloud Platform (GCP) for Geospatial Analysis

In this section, we will explore the application of the Google Cloud Platform (GCP) for geospatial analysis. GCP is a suite of cloud computing services offered by Google that provides a range of tools and services for storing, managing, and analyzing data in the cloud. It offers scalability, flexibility, and cost-effectiveness, making it an ideal platform for geospatial analysis tasks.

Background

Before diving into the specific features and tools provided by GCP for geospatial analysis, let's briefly review the fundamentals of geospatial analysis. Geospatial analysis is the process of analyzing and interpreting data that has a geographic or spatial component. It involves the use of spatial and statistical techniques to uncover patterns, relationships, and trends in data.

Geospatial analysis has many applications in various fields, including urban planning, environmental management, public health, and disaster management. Traditionally, geospatial analysis was conducted using desktop GIS software, which often required significant computational resources. With the advent of cloud computing, geospatial analysis can now be performed more efficiently and cost-effectively using cloud platforms like GCP.

Key Features of GCP for Geospatial Analysis

GCP offers several key features and services that are particularly useful for geospatial analysis tasks. Let's explore some of these features in detail:

Google Earth Engine

One of the most powerful tools provided by GCP for geospatial analysis is Google Earth Engine. It is a cloud-based platform that enables scientists, researchers, and developers to access a vast amount of geospatial imagery and geospatial data sets.

Google Earth Engine hosts a massive data catalog that includes satellite imagery, land cover data, climate data, and much more. Users can easily retrieve, analyze, and visualize these data sets using the Earth Engine API, which supports various programming languages, including Python and JavaScript.

With Earth Engine, you can perform complex geospatial analysis tasks, such as time series analysis, change detection, and land cover classification, at scale. The platform's powerful computing infrastructure allows for parallel processing, enabling the analysis of large volumes of geospatial data in a short amount of time.

BigQuery GIS

Another important feature of GCP for geospatial analysis is BigQuery GIS. BigQuery is a fully managed, serverless data warehouse provided by Google. It allows you to store, query, and analyze structured and semi-structured data at scale.

BigQuery GIS extends the capabilities of BigQuery to support geospatial data. It provides a set of SQL functions and data types for performing spatial queries and operations. You can store and analyze spatial data, such as points, lines, polygons, and rasters, in BigQuery tables.

By leveraging BigQuery GIS, you can perform spatial queries, such as finding points within a polygon or calculating the distance between two points, directly within the BigQuery environment. This integration of geospatial analysis with a powerful data warehousing solution makes BigQuery GIS a valuable tool for large-scale geospatial analysis tasks.

Cloud Storage and Compute Engine

GCP also offers reliable and scalable storage and compute resources that are essential for geospatial analysis. Cloud Storage provides a secure and durable object storage service for storing large volumes of geospatial data. It allows you to seamlessly integrate data stored in Cloud Storage with other GCP services, such as BigQuery and Earth Engine.

Compute Engine, on the other hand, provides virtual machine instances that you can use to process and analyze geospatial data. You can customize the virtual machine configurations to meet your specific requirements, such as computing power, memory, and storage.

By combining Cloud Storage and Compute Engine, you can store and process geospatial data in a scalable and efficient manner. This flexibility and scalability are particularly useful when working with large datasets or when performing computationally intensive geospatial analysis tasks.

Example: Flood Mapping Using GCP

To illustrate the application of GCP for geospatial analysis, let's consider an example of flood mapping. Flood mapping is a crucial task in disaster management, as it helps identify areas at risk of flooding, plan emergency response, and allocate resources effectively.

In this example, we will leverage the capabilities of GCP to analyze satellite imagery and generate flood maps for a specific region. Here are the steps involved:

1. Data Acquisition: Access satellite imagery for the target region from the Earth Engine data catalog or import your own data into Earth Engine. Ensure that the imagery covers the timeframe of interest, preferably before and after a flood event.

2. Preprocessing: Clean and preprocess the satellite imagery to remove noise, correct sensor artifacts, and enhance the visibility of water bodies. This step may involve image calibration, atmospheric correction, and image fusion techniques.

3. Image Classification: Apply supervised or unsupervised image classification techniques to distinguish water bodies from other land features in the imagery. This step helps generate a binary map indicating the presence or absence of water.

4. Change Detection: Compare the pre-flood and post-flood images to detect changes in water extent. This can be done using techniques such as image differencing or thresholding. The result is a map highlighting areas that have experienced flooding.

5. Spatial Analysis: Perform spatial analysis on the flood extent map to identify flood-prone areas, calculate flood frequency, and estimate the affected population or infrastructure. This can involve overlaying the flood map with other geographic data layers, such as population density or road networks, to assess the impact of flooding.

By leveraging the capabilities of GCP, such as Earth Engine for accessing satellite imagery, BigQuery GIS for spatial analysis, and Compute Engine for processing and storage, you can efficiently carry out the entire flood mapping workflow in a scalable and cost-effective manner.

Conclusion

In this section, we explored the application of the Google Cloud Platform (GCP) for geospatial analysis. GCP offers a range of tools and services that enable the storage, management, and analysis of geospatial data in the cloud. Features like Google Earth Engine, BigQuery GIS, and the storage and compute resources provided by GCP make it a powerful platform for conducting geospatial analysis tasks at scale.

By leveraging GCP, you can perform complex geospatial analysis tasks, such as time series analysis, change detection, and spatial queries, more efficiently and cost-effectively. The example of flood mapping illustrated how GCP can be used to generate flood maps and perform spatial analysis for disaster management purposes.

As cloud computing continues to evolve, it is becoming an indispensable tool for geospatial analysis professionals. GCP, with its rich set of geospatial analysis capabilities, is an excellent choice for those looking to leverage the power of the cloud in their geospatial analysis workflows.

Resources and Further Reading

To learn more about Google Cloud Platform (GCP) for geospatial analysis and related topics, consider exploring the following resources:

1. Google Cloud Platform documentation: https://cloud.google.com/docs/

2. Google Earth Engine documentation: https://earthengine.google.com/

3. BigQuery GIS documentation: https://cloud.google.com/bigquery/docs/gis-intro

4. "GeoSpatial Analysis: A Comprehensive Guide to Principles, Techniques, and Software Tools" by Michael J. de Smith, Michael F. Goodchild, and Paul A. Longley

5. "Python Geospatial Analysis Cookbook" by Michael Diener and Erik Westra

These resources provide in-depth information on various aspects of geospatial analysis, including cloud-based analysis platforms, spatial databases, and analytical techniques.

Microsoft Azure for Geospatial Analysis

In recent years, cloud computing has revolutionized the way data analysis is performed. It offers scalability, flexibility, and cost-effectiveness, making it an attractive option for geospatial analysis. In this section, we focus on the use of Microsoft Azure, a leading cloud computing platform, for geospatial analysis.

Introduction to Microsoft Azure

Microsoft Azure is a comprehensive cloud computing platform offered by Microsoft. It provides a wide range of services that enable organizations to build, deploy, and manage applications and services through Microsoft-managed data centers. Azure offers various tools and services for geospatial analysis, making it a popular choice among data analysts and geospatial professionals.

Azure Geospatial Services

Azure provides a rich set of geospatial services that facilitate effective geospatial analysis. Some of the key services are:

- **Azure Maps:** Azure Maps is a scalable and secure mapping service that offers geospatial analytics, visualization, and routing capabilities. It provides APIs and SDKs for developers to integrate maps and location-based services into their applications.
- **Azure Spatial Anchors:** Azure Spatial Anchors enables developers to build cross-platform augmented reality (AR) applications that use spatial mapping and spatial anchor capabilities. It allows for the creation of persistent and shared AR experiences that are spatially aligned across devices.
- **Azure Location Based Services:** Azure Location Based Services provides a suite of geographic API services that help developers create location-aware applications. It includes services for geocoding, routing, time zones, and traffic data, among others.
- **Azure IoT Hub:** Azure IoT Hub is a service that enables bi-directional communication between IoT devices and the

cloud. It allows for the ingestion, processing, and analysis of large volumes of geospatial data generated by IoT devices.

These services provide the foundation for performing advanced geospatial analysis tasks in the Azure environment.

Geospatial Data Storage in Azure

Azure offers various storage options for geospatial data, depending on the specific requirements of your analysis. Some of the key storage options are:

- **Azure Blob Storage:** Azure Blob Storage is a scalable and cost-effective storage solution for unstructured data. It is well-suited for storing large volumes of geospatial data such as satellite imagery, aerial photographs, and LiDAR datasets.

- **Azure Data Lake Storage:** Azure Data Lake Storage is a highly scalable and secure data lake that is optimized for big data analytics. It provides a hierarchical file system and supports the storage of structured, semi-structured, and unstructured data.

- **Azure SQL Database:** Azure SQL Database is a managed relational database service that offers high-performance and secure storage for structured geospatial data. It supports geospatial data types and provides built-in spatial functions for geospatial analysis.

With these storage options, you can effectively manage and analyze geospatial data in Azure.

Geospatial Analysis on Azure

Azure provides several tools and services for performing geospatial analysis on the cloud. Let's explore some of the key offerings:

- **Azure Machine Learning:** Azure Machine Learning is a cloud-based service that enables the development, deployment, and management of machine learning models. It provides a diverse set of tools and frameworks for geospatial machine learning tasks such as land cover classification, object detection, and spatial pattern recognition.

- **Azure Databricks:** Azure Databricks is a fast, easy, and collaborative Apache Spark-based analytics platform. It allows for distributed processing of geospatial data and supports a wide range of analytical tasks such as geospatial clustering, hotspot analysis, and geospatial regression modeling.

- **Azure HDInsight:** Azure HDInsight is a fully-managed cloud service that makes it easy to process big data using popular open-source frameworks such as Hadoop, Spark, and Hive. It enables geospatial analysts to perform large-scale geospatial data processing, geospatial data mining, and spatial data modeling.

These tools and services in Azure provide a comprehensive framework for conducting advanced geospatial analysis.

Case Study: Predictive Policing

To illustrate the capabilities of Microsoft Azure for geospatial analysis, let's consider a case study on predictive policing. Predictive policing is the use of analytical techniques to identify potential crime hotspots and allocate law enforcement resources accordingly.

In this case study, we can leverage Azure Machine Learning and Azure Databricks to perform geospatial analysis on historical crime data. By integrating various geospatial datasets such as crime incidents, demographic information, and environmental factors, we can build machine learning models to predict the likelihood of crime occurrence in specific locations.

Azure Machine Learning provides a range of algorithms and tools for modeling and training predictive policing models. By utilizing geospatial features such as crime density, proximity to known hotspots, and socio-economic indicators, we can develop accurate predictive models that inform law enforcement agencies about potential crime hotspots.

Azure Databricks comes into play for the processing of large-scale geospatial data and the execution of complex geospatial analysis algorithms. It allows for the integration of spatial data sources and the exploration of spatial patterns and relationships. With its distributed computing capabilities, Azure Databricks enables efficient geospatial analysis at scale.

By combining the power of Azure Machine Learning and Azure Databricks, law enforcement agencies can make data-driven decisions and allocate resources effectively to prevent crime.

Conclusion

Microsoft Azure offers a powerful and comprehensive platform for geospatial analysis. With its rich set of geospatial services, storage options, and analytical tools, Azure provides a flexible and scalable environment for conducting advanced geospatial analysis tasks. Whether it is machine learning, big data processing, or augmented reality applications, Azure can handle the complexities of geospatial analysis in the cloud. By leveraging the capabilities of Azure, organizations can unlock new insights from their geospatial data and drive data-informed decision-making.

Resources and Further Reading

To learn more about using Microsoft Azure for geospatial analysis, here are some recommended resources:

- Microsoft Azure Documentation: `https://docs.microsoft.com/azure/`
- "Mastering Geospatial Analysis with Python" by Silas Toms and Paul Crickard IV
- "Big Data Analytics with Azure Machine Learning" by Richard Conway and Ryan Dunn
- "Geospatial Analysis - A Comprehensive Guide" by Michael J. de Smith, Michael F. Goodchild, and Paul A. Longley

These resources provide in-depth information and practical examples of geospatial analysis using Microsoft Azure.

Geospatial Data Analysis in Real-World Applications

Urban Planning and Smart Cities

Urban planning plays a crucial role in shaping the development of cities and communities. With the advent of new technologies and the concept of smart cities, the field of urban planning has been revolutionized. Smart cities are characterized by the integration of information and communication technologies (ICT) to enhance the quality of life, improve sustainability, and streamline urban services. In this section, we will explore the application of geospatial data analysis and mapping in urban planning and smart cities.

Challenges in Urban Planning

Urban planning faces numerous challenges, such as population growth, resource management, transportation, and infrastructure development. Geospatial data analysis provides valuable insights into these challenges by harnessing the power of data and spatial analysis techniques.

One of the key challenges in urban planning is land use optimization. Given limited land resources, it is essential to allocate land for different purposes effectively. Geospatial data analysis helps identify the suitable locations for residential, commercial, and industrial purposes based on factors like accessibility, infrastructure availability, and environmental considerations.

Another challenge is transportation planning. Geospatial data analysis enables the analysis of transportation networks, traffic patterns, and travel demand to optimize transportation systems. It helps identify areas with high congestion, plan new infrastructure projects, and optimize public transportation routes.

Additionally, urban planning requires a holistic view of sustainability. Geospatial data analysis can assess the environmental impacts of urban development, analyze energy consumption patterns, and identify opportunities for renewable energy integration. It also helps analyze waste management, water resources, and green space distribution for sustainable urban development.

Role of Geospatial Data Analysis

Geospatial data analysis plays a pivotal role in urban planning and smart cities. It provides a spatial perspective to analyze and visualize complex urban systems. By incorporating various data sources, such as satellite imagery, sensor data, and socioeconomic data, geospatial analysis enables planners to make informed decisions.

One of the fundamental applications of geospatial data analysis in urban planning is land-use planning. By analyzing land cover data, zoning regulations, and demographic information, planners can identify suitable areas for different land uses. Geospatial analysis facilitates the identification of land parcels that adhere to specific criteria, such as proximity to amenities, infrastructure availability, and environmental constraints.

Transportation planning is another area where geospatial data analysis is indispensable. By analyzing transportation networks, traffic patterns, and travel demand, planners can optimize transportation systems, reduce congestion, and improve accessibility. Geospatial analysis helps in identifying areas with inadequate transportation facilities and planning new infrastructure to address these gaps.

Furthermore, geospatial data analysis assists in urban environmental management. It enables the assessment of air quality, noise pollution, and green space distribution. By integrating sensor data with geographical information, planners can monitor and manage environmental parameters in real-time. Geospatial analysis also aids in identifying vulnerable areas to natural disasters and implementing preventive measures.

Data Sources and Tools

To perform geospatial data analysis in urban planning, several data sources and tools are available. Satellite imagery, aerial photography, and remote sensing data provide high-resolution spatial information about land cover, vegetation patterns, and urban growth. Open-source geospatial datasets, such as OpenStreetMap, provide comprehensive data about transportation networks, amenities, and points of interest.

Geographic Information Systems (GIS) software forms the backbone of geospatial data analysis in urban planning. Tools like QGIS, ArcGIS, and Python libraries like GeoPandas and Rasterio facilitate spatial data manipulation, analysis, and visualization. These software tools offer a wide array of spatial analysis techniques, including buffering, overlay analysis, and network routing.

In recent years, the emergence of cloud computing platforms and APIs has made geospatial analysis more accessible. Platforms like Google Earth Engine and Amazon Web Services provide scalable computing power and extensive geospatial datasets for analysis. These platforms enable the processing of large-scale geospatial data and the implementation of advanced analysis algorithms.

Visualization and Communication

Effective visualization and communication of geospatial data analysis results are crucial for urban planning and smart cities. Data visualization techniques help planners and stakeholders understand complex spatial relationships and patterns. Interactive maps, charts, and graphs improve the accessibility and clarity of analysis results.

GIS software, such as QGIS and ArcGIS, offer a range of visualization options to create compelling maps. Python libraries like Matplotlib, Plotly, and Folium provide additional tools for data visualization and interactive mapping. These libraries enable the creation of thematic maps, time animation, and interactive web maps to communicate the findings effectively.

In addition to traditional visualization techniques, emerging technologies like virtual reality (VR) and augmented reality (AR) offer new possibilities for urban planning visualization. VR and AR can provide immersive experiences, allowing planners and stakeholders to visualize planned developments in real-world contexts. These technologies enhance stakeholder engagement and decision-making processes.

Case Study: Optimal Location for a Smart Park

To illustrate the application of geospatial data analysis in urban planning, let's consider a case study of locating a smart park in a city. A smart park integrates technology and smart infrastructure to enhance visitor experience and improve sustainability.

First, we need to analyze various factors like accessibility, existing green spaces, and population density. Using GIS software and geospatial data, we can identify areas with low accessibility to existing parks and high population density. These areas present a potential opportunity to establish a smart park.

Next, we can perform a suitability analysis by considering factors like proximity to residential areas, availability of amenities, and environmental constraints. By overlaying various datasets, we can assign weights to different factors and generate a suitability map. This map helps identify the optimal location for the smart park based on the predefined criteria.

Once the location is determined, geospatial data analysis can assist in designing the park layout, incorporating smart technologies, and optimizing the park's energy consumption. By considering factors like solar potential, pedestrian flow, and noise pollution, planners can create a sustainable and interactive environment.

Conclusion

Geospatial data analysis and mapping play a vital role in urban planning and the development of smart cities. By leveraging the power of data, spatial analysis techniques, and visualization tools, planners can gain valuable insights into land use, transportation, sustainability, and environmental management. The integration of geospatial data analysis in decision-making processes enhances the effectiveness and efficiency of urban planning, leading to better-designed cities that are more sustainable and livable. As technologies continue to advance, the role of geospatial data analysis will become even more critical in shaping the cities of the future.

Resources and Further Reading

For readers interested in delving deeper into the subject of geospatial data analysis, urban planning, and smart cities, the following resources are highly recommended:

- Books:
  - "GIS and Geocomputation for Water Resource Science and Engineering" by Barnali Dixon
  - "Smart Cities: Big Data, Civic Hackers, and the Quest for a New Utopia" by Anthony M. Townsend
  - "GIS for Environmental Applications: A Practical Approach" by Xuan Zhu and Jiulin Sun
- Online Courses:
  - Coursera: "Geographic Information Systems (GIS)" by University of California, Davis
  - edX: "Smart Cities, Management of Smart Urban Infrastructures" by École Polytechnique Fédérale de Lausanne
  - Esri Training: Various courses on GIS, geospatial analysis, and urban planning
- Websites:
  - Urban Land Institute (ULI): `https://www.uli.org`
  - Smart Cities Council: `https://smartcitiescouncil.com`
  - ESRI GIS for Urban and Regional Planning: `https://www.esri.com/en-us/arcgis/smart-mapping/urban-planning`

These resources provide a comprehensive understanding of geospatial data analysis, urban planning, and smart cities, enabling you to gain deeper insights into these fascinating fields.

Public Health and Epidemiology

Public health and epidemiology are two interconnected fields that aim to improve the health of populations through the study of diseases, health behaviors, and risk factors. In this section, we will explore how geospatial data analysis can enhance our understanding of public health and epidemiological patterns, and how Python can be used to analyze and visualize such data.

Introduction to Public Health and Epidemiology

Public health focuses on the prevention and promotion of health at the population level. It encompasses a wide range of activities, from surveillance and monitoring of health outcomes to policy development and health promotion.

Epidemiology, on the other hand, is the study of the distribution and determinants of diseases in populations. It involves investigating the causes and risk factors associated with diseases, as well as developing strategies for disease prevention and control.

Both public health and epidemiology rely heavily on data analysis to identify patterns, trends, and associations between various factors and health outcomes. Geospatial data analysis plays a crucial role in understanding the spatial distribution of diseases, identifying high-risk areas, and planning effective interventions.

Geospatial Analysis in Public Health

Geospatial analysis in public health involves the integration of geographic information systems (GIS) and epidemiological methods to investigate the spatial patterns and determinants of health outcomes. By analyzing geospatial data, public health researchers and practitioners can:

- Identify clusters or hotspots of diseases: Geospatial analysis allows for the detection of clusters or areas with a higher than expected number of cases. This information helps in understanding the spatial distribution of diseases and can assist in targeting interventions and resources to specific areas.

- Assess environmental risk factors: Geospatial analysis enables the investigation of the relationship between environmental exposures and health outcomes. For example, it can be used to study the impact of air pollution on respiratory diseases or the relationship between neighborhood characteristics and cardiovascular health.

- Model disease spread and transmission: Geospatial analysis can be used to develop models that simulate the spread and transmission of diseases. These models help in predicting disease outbreaks, evaluating the impact of interventions, and optimizing resource allocation.

Geospatial Data Collection in Public Health

Geospatial data used in public health research can be obtained from various sources, including:

- Health surveys and registries: Data from surveys, registries, and health records provide information on disease prevalence, incidence, and demographic characteristics of affected individuals. These data can be geocoded to enable spatial analysis.

- Remote sensing data: Satellite imagery and remote sensing data provide valuable information on environmental factors that may influence health outcomes. Examples include land cover, temperature, humidity, and air pollution measurements.

- Social media and mobile data: Data from social media platforms and mobile devices offer insights into health-related behaviors, such as physical activity patterns, food choices, and exposure to health risks. These data sources can be particularly useful for tracking disease outbreaks or studying the impact of interventions.

Geospatial Analysis Techniques

Various geospatial analysis techniques can be applied to public health and epidemiological data. Some commonly used techniques include:

- Spatial clustering analysis: This technique helps in identifying clusters or hotspots of disease occurrences. It uses statistical methods to determine whether the observed spatial pattern is due to random chance or represents a true cluster.

- Spatial interpolation: Interpolation methods, such as kriging, can be used to estimate health-related variables at unmeasured locations based on known data points. This is particularly useful when spatially continuous information is needed, such as creating health risk maps.

- Network analysis: Network analysis techniques, such as distance decay analysis and network centrality measures, can be applied to study the spread of diseases through transport networks or social networks.

- Geographically weighted regression: Geographically weighted regression allows for the investigation of spatially varying relationships between predictors and health outcomes. It helps in understanding how the effects of risk factors may vary across different locations.

Case Study: Mapping Disease Outbreaks

Let's consider a case study of mapping disease outbreaks using geospatial analysis techniques. Suppose there has been a recent outbreak of a mosquito-borne disease in a city, and the public health department wants to understand the spatial distribution of cases to inform targeted interventions.

To analyze this, the public health department collects data on confirmed cases, including their geographical coordinates. Using Python and geospatial libraries such as GeoPandas and Folium, we can visualize the distribution of cases on a map. We can also conduct spatial clustering analysis to identify areas with a high concentration of cases.

Furthermore, additional data on environmental factors, such as temperature and rainfall, can be incorporated to assess their potential influence on the disease spread. This can be done through spatial regression analysis, where we model the relationship between disease cases and environmental variables.

By combining information from multiple sources and applying geospatial analysis techniques, the public health department can gain valuable insights into the patterns and determinants of the disease outbreak. This information can assist in resource allocation, targeting interventions, and implementing preventive measures.

Ethical Considerations

While geospatial data analysis offers numerous opportunities for public health and epidemiology, it also raises ethical considerations. It is essential to ensure privacy and confidentiality when collecting and analyzing geospatial data, especially when dealing with individual-level health data.

Additionally, the interpretation and dissemination of geospatial analysis results should be done carefully to avoid stigmatizing specific areas or communities. Efforts should be made to engage with local communities and ensure that interventions are culturally appropriate and respectful of local contexts.

Summary

Geospatial data analysis plays a critical role in public health and epidemiology by enhancing our understanding of disease patterns, identifying high-risk areas, and informing targeted interventions. Through the integration of geographic information systems and epidemiological methods, we can assess the impact of environmental factors, model disease spread, and analyze spatial relationships.

In this section, we explored the principles of public health and epidemiology, discussed the applications of geospatial analysis in these fields, and highlighted various geospatial analysis techniques. We also presented a case study illustrating the use of geospatial analysis to map disease outbreaks.

Geospatial data analysis in public health is a rapidly evolving field, with new technologies and data sources constantly emerging. As researchers and practitioners, it is essential to stay updated with advancements in geospatial analysis and leverage the power of Python and geospatial libraries to address public health challenges and improve population health.

Natural Hazards and Disaster Management

Natural hazards and disasters pose significant threats to human lives and property. In recent years, the field of geospatial data analysis has played a crucial role in understanding and mitigating these risks. This section will explore the application of advanced techniques in geospatial data analysis for natural hazards and disaster management.

Introduction to Natural Hazards

Natural hazards refer to extreme natural events that can cause widespread destruction and have a significant impact on human populations. Examples of natural hazards include earthquakes, floods, hurricanes, wildfires, landslides, and volcanic eruptions. These events can result in the loss of lives, displacement of people, damage to infrastructure, and disruption of essential services.

Understanding the key characteristics of natural hazards is essential for effective disaster management. Geospatial data analysis provides valuable insights into hazard identification, vulnerability assessment, and risk analysis.

Geospatial Data Acquisition for Natural Hazards Analysis

Accurate and up-to-date geospatial data is essential for assessing and mitigating the impact of natural hazards. There are various sources of geospatial data that can be used for hazard analysis, including remote sensing, satellite imagery, aerial photography, and on-ground measurements.

Remote sensing platforms, such as satellites and drones, capture high-resolution imagery that can be used to detect changes in the environment. This data can help identify areas prone to natural hazards, monitor their evolution, and assess their impact after an event. Additionally, ground-based sensors and monitoring networks provide real-time data on weather conditions, seismic activity, and other relevant parameters.

2. Data Cleaning and Quality Assurance

Before geospatial data can be used for hazard analysis, it is crucial to clean and preprocess the data to ensure its quality and reliability. This process involves removing outliers, correcting errors, and integrating data from different sources.

Data cleaning techniques for geospatial data include outlier detection, filtering, and interpolation methods. Quality assurance measures, such as accuracy assessment, error estimation, and uncertainty analysis, are also essential to evaluate the reliability of the data.

3. Geospatial Data Analysis Techniques for Natural Hazards

Geospatial data analysis techniques play a vital role in understanding the characteristics and behaviors of natural hazards. These techniques enable the identification of patterns, trends, and relationships within the data, which can inform hazard assessment and management strategies.

a. Spatial Analysis: Spatial analysis techniques, such as spatial clustering and hotspots analysis, can help identify areas with higher vulnerability or frequency of natural hazards. These techniques also enable the identification of spatial patterns and relationships between hazard events and environmental factors.

b. Time Series Analysis: Time series analysis techniques can be used to analyze the temporal evolution of natural hazards, such as the frequency and intensity of earthquakes or the precipitation patterns leading to floods. These analyses can provide valuable information for hazard forecasting and early warning systems.

c. Machine Learning: Machine learning algorithms, such as random forests and support vector machines, can be applied to geospatial data for hazard prediction and classification. These techniques enable the development of predictive models based on historical data, environmental parameters, and other relevant factors.

4. Risk Assessment and Mitigation Strategies

Geospatial data analysis plays a critical role in risk assessment and the development of effective mitigation strategies for natural hazards. By integrating geospatial data with demographic, socio-economic, and infrastructure information, risk models can be developed to identify high-risk areas and vulnerable populations.

Risk assessment techniques, such as probabilistic modeling and multi-criteria decision analysis, can help quantify the likelihood and impacts of natural hazard events. This information can guide the allocation of resources for hazard mitigation and emergency response planning.

5. Case Studies and Real-World Applications

To illustrate the practical application of geospatial data analysis in natural hazards and disaster management, let's consider a real-world case study: flood risk assessment in a coastal city.

In this scenario, geospatial data, including elevation models, land cover data, and historical flood records, are used to develop a flood risk map for the city. Geospatial analysis techniques, such as hydrological modeling and flood inundation mapping, are applied to identify areas at high risk of flooding.

The flood risk map can inform urban planning and land-use decisions, aiding in the development of flood-resilient infrastructure and evacuation plans. Furthermore, real-time monitoring of weather conditions and river levels can be integrated into the analysis, enabling timely flood forecasting and early warning systems.

6. Challenges and Ethical Considerations in Geospatial Data Analysis

While geospatial data analysis offers immense potential for natural hazards and disaster management, there are several challenges and ethical considerations to be aware of.

a. Data Accessibility and Privacy: Access to high-quality geospatial data can be challenging, particularly in developing countries or areas with limited resources. Additionally, privacy concerns arise when utilizing sensitive data, such as population demographics or infrastructure locations, in hazard analysis.

b. Uncertainty and Modeling Limitations: Geospatial data analysis involves uncertainties, such as errors in data sources or limitations in modeling techniques. It is crucial to acknowledge and communicate these uncertainties to ensure the appropriate interpretation and use of the analysis results.

c. Interdisciplinary Collaboration: Geospatial data analysis for natural hazards requires collaboration between various disciplines, including geography, geology, meteorology, and urban planning. Effective communication and coordination among experts are critical to ensure comprehensive and accurate hazard assessments.

7. Resources and Further Reading

To further explore the applications of geospatial data analysis in natural hazards and disaster management, the following resources are recommended:

- Geospatial Analysis - A Comprehensive Guide by Michael J. De Smith, et al. - Geographic Information Systems and Science by Paul A. Longley, et al. - Remote Sensing and GIS in Disaster Management edited by U. M. Chandrashekhar and S. R. Gopala Krishna

Additionally, online resources, such as the United Nations Office for Disaster Risk Reduction (UNDRR) and the International Federation of Red Cross and Red Crescent

Societies (IFRC), provide valuable information on disaster risk reduction strategies and best practices.

In conclusion, geospatial data analysis plays a vital role in natural hazards and disaster management. By harnessing the power of geospatial data and advanced analytical techniques, decision-makers can better understand, predict, and mitigate the impact of natural hazards on human lives and infrastructure. Through interdisciplinary collaboration and ethical considerations, geospatial data analysis can contribute to building more resilient and safer communities.

Conclusion

Summary of Key Concepts

In this section, we will summarize the key concepts that have been covered throughout this textbook on advanced techniques in geospatial data analysis and mapping with Python. We will review the main ideas, principles, and techniques that are essential for understanding and applying geospatial data analysis in various domains.

Background of Geospatial Data Analysis

Geospatial data analysis is the process of exploring, analyzing, and visualizing spatial data to uncover patterns, relationships, and trends in the geographical world around us. It involves the use of geographic information systems (GIS) and spatial statistical methods to understand the spatial distribution of phenomena and their underlying processes. Geospatial data analysis is widely used in fields such as urban planning, environmental management, epidemiology, and disaster management.

Data Acquisition and Preprocessing

The first step in any geospatial analysis is to acquire and preprocess the data. This involves identifying and obtaining relevant datasets from various sources, such as remote sensing, GPS, and demographic data. Data cleaning and quality assurance techniques are then applied to address data errors, missing values, and outliers. Data integration and interoperability methods ensure compatibility between different data formats and standards.

Exploratory Data Analysis

Exploratory data analysis (EDA) is a crucial step in geospatial data analysis. Descriptive statistics and data visualization techniques help us understand the basic properties and patterns of the data. Spatial autocorrelation analysis measures the degree of similarity or dissimilarity between spatial observations. Spatial clustering techniques identify groups or clusters of similar spatial objects. Geospatial data mining techniques extract useful knowledge from large spatial datasets.

Geostatistical Analysis

Geostatistical analysis is a powerful set of methods used to model and analyze spatial data. It provides tools for estimation, interpolation, and prediction of unknown values at unobserved locations. Ordinary kriging is a widely used geostatistical method that produces unbiased estimates of missing values based on spatial dependence. Co-kriging and co-regression techniques incorporate auxiliary variables to improve prediction accuracy. Bayesian geostatistics allows for the integration of prior knowledge and uncertainty in the analysis.

Spatial Data Visualization and Cartography

Spatial data visualization plays a crucial role in geospatial data analysis and mapping. Principles of data visualization guide us in creating effective and informative visual representations of spatial data. Python libraries such as Matplotlib, Seaborn, and Plotly provide powerful tools for mapping and visual exploration. Advanced data visualization techniques, such as 3D mapping, heatmaps, and interactive dashboards, enhance the understanding and communication of spatial patterns. Cartographic design and layout principles help us create aesthetic maps that effectively convey information.

Spatial Regression and Machine Learning

Spatial regression and machine learning methods are employed to model and analyze the relationships between spatial variables. Spatial regression techniques extend traditional regression models to account for spatial dependencies in the data. Machine learning algorithms, such as random forests and support vector machines, are applied to geospatial data analysis for classification and prediction tasks. Spatial econometric models combine both spatial regression and machine learning approaches to analyze spatially dependent data.

Geospatial Network Analysis

Geospatial network analysis focuses on the analysis and modeling of spatially interconnected networks, such as transportation networks, social networks, and utility networks. Introduction to geospatial networks covers basic concepts and terminologies related to networks. Network analysis in Python introduces tools and algorithms for route optimization, network connectivity, and centrality analysis. Spatial network analysis techniques allow for the identification of critical network components, route planning, and facility location optimization.

Geospatial Data Analysis in the Cloud

The use of cloud computing platforms for geospatial data analysis provides scalability, flexibility, and cost-effectiveness. Introduction to cloud computing and GIS explains the benefits and challenges of cloud-based geospatial analysis. Amazon Web Services (AWS), Google Cloud Platform (GCP), and Microsoft Azure are popular cloud platforms that offer geospatial analysis services. Understanding cloud-based solutions helps researchers and practitioners leverage the power of distributed computing for large-scale geospatial data analysis.

Geospatial Data Analysis in Real-World Applications

Geospatial data analysis has wide-ranging applications in real-world scenarios. Urban planning and smart cities utilize geospatial analysis to optimize land use, transportation systems, and city infrastructure. Environmental management and sustainability employ geospatial tools to monitor and protect natural resources. Public health and epidemiology benefit from geospatial analysis in disease surveillance, risk assessment, and resource allocation. Natural hazards and disaster management apply geospatial analysis to assess vulnerability, plan emergency responses, and aid in recovery efforts.

Conclusion

In conclusion, this textbook has covered a wide range of advanced techniques in geospatial data analysis and mapping using Python. We have explored the fundamental concepts, methods, and tools in acquiring, preprocessing, analyzing, and visualizing geospatial data. We have also examined various applications of geospatial analysis in real-world domains. By mastering these key concepts, readers will be well-equipped to tackle complex geospatial data analysis challenges and contribute to the advancement of knowledge and decision-making in their respective fields.

For more information and further reading on the topics covered in this textbook, the following resources are recommended:

- Goodchild, M. F., Janelle, D. G., & Hackett, S. (Eds.). (2010). *Geospatial Analysis: A Comprehensive Guide to Principles, Techniques and Software Tools.*

- Longley, P. A., Goodchild, M. F., Maguire, D. J., & Rhind, D. W. (2015). *Geographic Information Science & Systems.*

- O'Sullivan, D., & Unwin, D. J. (2018). *Geographic Information Analysis.*

- Rogerson, P. A. (2019). *Statistical Methods for Geography: A Student's Guide.*

- Rushton, G., Armstrong, M. P., Gittler, J., Greene, B. R., & Pavlik, C. E. (2020). *Geocoding Health Data: The Use of Geographic Codes in Cancer Prevention and Control, Research and Practice.*

Remember, geospatial data analysis is a rapidly evolving field, so it is essential to keep up to date with current research, emerging technologies, and open-source software tools. Stay curious, seek new challenges, and continue exploring the fascinating world of geospatial data analysis!

Future Trends and Emerging Technologies

In the world of geospatial data analysis and mapping, there are always new trends and technologies emerging that shape the future of the field. These advancements open up exciting opportunities to explore and analyze spatial data in innovative ways. In this section, we will discuss some of the future trends and emerging technologies that are expected to have a significant impact on geospatial data analysis.

Artificial Intelligence and Machine Learning in Geospatial Analysis

One of the most prominent trends in geospatial data analysis is the integration of artificial intelligence (AI) and machine learning (ML) techniques. AI and ML algorithms have the potential to revolutionize geospatial analysis by automating processes, improving accuracy, and uncovering hidden patterns in spatial data.

AI and ML techniques can be applied to a wide range of geospatial tasks, including image classification, object detection, feature extraction, and prediction modeling. For example, in satellite imagery analysis, AI algorithms can automatically detect and classify land cover types, identify changes in vegetation patterns, and monitor urban expansion.

Moreover, ML algorithms can be used to analyze and predict natural disasters such as floods and wildfires, enabling timely response and disaster management planning. By training these algorithms on historical data, they can learn the complex relationships between various environmental factors and predict the likelihood of future events.

Big Data Analytics and Cloud Computing

With the exponential growth of geospatial data, handling and analyzing massive datasets has become a significant challenge. This has led to the emergence of big data analytics techniques and cloud computing solutions specifically tailored for geospatial data analysis.

Big data analytics frameworks, such as Apache Hadoop and Apache Spark, provide scalable and distributed processing capabilities that can handle large volumes of spatial data. These frameworks allow analysts to perform complex geospatial queries, spatial aggregations, and spatial join operations on massive datasets efficiently.

In addition, cloud computing platforms, such as Amazon Web Services (AWS), Google Cloud Platform (GCP), and Microsoft Azure, offer geospatial services and infrastructure to store, process, and visualize spatial data. These cloud-based solutions provide on-demand computing resources, allowing analysts to scale their geospatial analysis tasks based on their requirements.

Internet of Things (IoT) and Sensor Networks

The Internet of Things (IoT) has opened up new avenues for collecting and analyzing geospatial data. IoT devices and sensor networks are generating vast amounts of spatially-referenced information in real-time, enabling a wide range of applications in various domains.

For example, in the field of transportation, IoT devices and sensors embedded in vehicles, traffic lights, and roads can collect data on traffic flow, congestion, and parking availability. This real-time data can be analyzed to optimize transportation systems, improve traffic management, and provide accurate travel time estimations.

Similarly, in agriculture, IoT devices and sensors deployed in fields can monitor soil moisture levels, temperature, and crop health. This data can be analyzed to optimize irrigation schedules, detect disease outbreaks, and improve crop yield.

Spatial Data Science and Open-Source Tools

As the complexity of geospatial data analysis continues to evolve, there is a growing need for spatial data scientists who possess both domain-specific knowledge and data science skills. Spatial data science combines the principles of geospatial analysis with statistical modeling, data visualization, and machine learning techniques.

Open-source tools and libraries, such as Python's GeoPandas, PySAL, and scikit-learn, are playing a crucial role in democratizing spatial data science. These tools provide a wide range of functionalities for geospatial analysis, including spatial data manipulation, exploratory data analysis, and modeling.

By leveraging open-source tools, spatial data scientists can harness the power of these libraries and contribute to the development of innovative geospatial analysis techniques. The open nature of these tools also promotes collaboration and knowledge sharing within the geospatial community.

Challenges and Ethical Considerations

While the future trends and emerging technologies in geospatial data analysis offer immense potential, they also raise several challenges and ethical considerations.

One of the challenges is the issue of data privacy and security. As geospatial data becomes more accessible and widely available, protecting the privacy of individuals and sensitive locations becomes paramount. It is crucial to develop robust frameworks and guidelines for handling and sharing geospatial data while complying with privacy regulations.

Another challenge is the potential bias in AI and ML algorithms. These algorithms are trained on historical data, which can reflect societal biases and disparities. It is essential to mitigate these biases and ensure that the algorithms used in geospatial analysis are fair, transparent, and accountable.

Furthermore, the increasing reliance on cloud computing raises concerns about data sovereignty and vendor lock-in. Organizations must carefully evaluate the potential risks associated with storing sensitive geospatial data on third-party cloud platforms and consider strategies to mitigate these risks.

Conclusion

The future of geospatial data analysis and mapping is exciting, with emerging technologies revolutionizing the way we analyze and visualize spatial information. Artificial intelligence and machine learning, big data analytics, Internet of Things, and open-source tools are among the key trends shaping the field.

However, it is essential to tackle the challenges and ethical considerations associated with these technologies to ensure responsible and equitable geospatial data analysis practices. By addressing these issues, we can unlock the full potential of geospatial data analysis and make informed decisions for a better future.

Resources and Further Reading

- Goodchild, M. F., & Janelle, D. G. (Eds.). (2010). Spatially integrated social science. Oxford University Press.

- Longley, P. A., Goodchild, M. F., Maguire, D. J., & Rhind, D. W. (2018). Geographic Information Science and Systems. John Wiley & Sons.

- Miller, H. J., & Han, J. (2009). Geographic Data Mining and Knowledge Discovery. CRC Press.

- O'Sullivan, D., & Unwin, D. J. (2018). Geographic Information Analysis. John Wiley & Sons.

Case Studies and Real-World Applications

In this section, we will explore several case studies and real-world applications of geospatial data analysis using Python. These examples will demonstrate the practical applications of the techniques and concepts discussed throughout the book.

Case Study 1: Urban Planning and Transportation

Urban planning plays a crucial role in shaping the development and growth of cities. Geospatial data analysis can provide valuable insights for urban planners to enhance transportation systems, optimize land use, and improve overall livability. Let's consider a case study focused on analyzing transportation patterns in a metropolitan area to identify areas of congestion and propose potential solutions.

Problem Statement

The city of Metropolis is experiencing increased traffic congestion, leading to longer commute times, environmental pollution, and decreased quality of life. The city planners want to analyze the traffic flow data collected from various sensors placed across the city and identify the key areas responsible for congestion. They also want to propose alternative routes or transportation solutions to alleviate the traffic burden.

Solution

To address this problem, the city planners can leverage geospatial data analysis techniques. They can start by acquiring and pre-processing traffic data collected from sensors using Python. By applying exploratory data analysis techniques, such as descriptive statistics and data visualization, they can analyze the traffic patterns, identify congested areas, and understand the factors contributing to congestion.

Next, the planners can utilize geostatistical analysis techniques, such as spatial autocorrelation analysis and spatial clustering, to identify hotspots of congestion and categorize the city areas based on traffic density. This analysis will provide valuable insights into the spatial distribution of congestion and help prioritize areas for intervention.

With the knowledge gained from the geospatial analysis, the planners can propose alternative transportation routes or suggest improvements to existing infrastructure. They can also leverage spatial regression and machine learning techniques to model and predict traffic flow under different scenarios, facilitating better decision-making for urban transportation planning.

Challenges and Considerations

While analyzing urban transportation data, various challenges may arise. One challenge is the high volume and complexity of geospatial data, which requires efficient data preprocessing and handling techniques. Additionally, ensuring data quality and accuracy is crucial for obtaining reliable results. Another challenge is the dynamic nature of urban environments, where traffic patterns, infrastructure, and population densities change over time. Therefore, it is essential to update and validate geospatial models regularly.

Ethical considerations should also be taken into account when performing geospatial analysis for urban planning. Privacy concerns may arise when dealing with sensitive personal data collected through sensors or other sources. Proper anonymization and aggregation techniques must be implemented to protect individual privacy.

Case Study Results

After analyzing the traffic data in Metropolis, the city planners identified key areas of congestion, such as major intersections and highways. They discovered that the increase in population density and the lack of efficient public transportation options contributed to the problem. Based on the results, the planners recommended implementing dedicated bus lanes, improving intermodal transportation options, and implementing traffic signal optimization strategies to alleviate congestion and improve traffic flow.

Case Study 2: Environmental Management and Conservation

Geospatial data analysis has proven to be invaluable in environmental management and conservation efforts. By utilizing various geospatial analysis techniques, researchers and conservationists can monitor and assess the impact of environmental changes, track endangered species, and design effective conservation strategies. Let's explore a case study focused on biodiversity conservation and habitat mapping.

Problem Statement

A team of ecologists is working on a conservation project in a national park known for its rich biodiversity. They aim to identify and map critical habitats for endangered species within the park boundaries. The availability of such detailed habitat maps will aid in designing conservation strategies, setting up protected zones, and monitoring changes over time.

Solution

To address this problem, the team can employ geospatial data analysis techniques using Python. They can start by acquiring satellite imagery and other relevant geospatial data, such as vegetation indices, elevation data, and climate variables. By integrating and preprocessing this data, they can create a comprehensive spatial database for analysis.

Next, they can utilize exploratory data analysis techniques to identify areas with high biodiversity and ecological significance. They can calculate various indices, such as species richness and evenness, and visualize the spatial distribution of species using heatmaps and choropleth maps.

To map critical habitats, the team can employ geostatistical analysis techniques like ordinary kriging or co-kriging to predict species distributions based on environmental variables. They can also leverage machine learning algorithms to model habitat preferences and predict suitable habitat patches for endangered species.

Challenges and Considerations

When dealing with environmental data analysis, several challenges may arise. One significant challenge is the availability and quality of data. Remote sensing data, for example, may have limitations in terms of resolution and accuracy. Additionally, the spatial and temporal dynamics of ecological systems can pose challenges for modeling and analysis.

Another consideration is the trade-off between conservation efforts and human activities. Different stakeholders may have conflicting objectives, making it essential to balance conservation goals with socioeconomic factors. Ethical considerations also come into play when monitoring or tracking endangered species, where stress and disturbance to wildlife should be minimized.

Case Study Results

The team successfully mapped critical habitats for several endangered species within the national park. They identified key areas for habitat conservation and proposed the establishment of protected zones to minimize human disturbances. The habitat maps and conservation strategies guided the park management in their efforts to ensure the long-term survival of the endangered species and maintain the park's ecological balance.

Additional Resources

For further exploration of case studies and real-world applications in geospatial data analysis, the following resources are recommended:

- Goodchild, M. F., & Janelle, D. G. (Eds.). (2010). *Spatially Integrated Social Science.* Oxford University Press.
- Geospatial Analysis - A comprehensive guidebook to spatial analysis techniques and GIS (https://geospatial-analysis.com/)
- Python Geospatial Development - Effective techniques for geospatial development using Python (https://www.packtpub.com/application-development/python-g
- Environmental Modeling and GIS - Spatial analysis and decision support (https://www.springer.com/gp/book/9780387952856)

With the knowledge gained from the examples and resources provided in this section, you can now confidently apply geospatial data analysis techniques to diverse real-world problems. The field of geospatial analysis continues to grow, and it offers endless opportunities for innovation and solving complex spatial challenges.

Challenges and Ethical Considerations in Geospatial Data Analysis

Geospatial data analysis involves the use of various techniques to extract information and insights from spatial data. However, this field also presents certain challenges and ethical considerations that need to be addressed. In this section, we will discuss some of the key challenges and ethical considerations in geospatial data analysis.

Data Quality and Accuracy

One of the primary challenges in geospatial data analysis is ensuring the quality and accuracy of the data. Spatial data can come from multiple sources and may suffer from issues such as errors, inconsistencies, or outdated information. These issues can significantly impact the results and conclusions drawn from the analysis.

To overcome this challenge, it is crucial to implement robust data cleaning and quality assurance processes. This involves identifying and correcting errors, resolving inconsistencies, and validating the accuracy of the data. Additionally, thorough documentation and metadata management are essential to ensure transparency and reproducibility in geospatial analysis.

Data Privacy and Security

Geospatial data often contains sensitive information, such as personal or location-based data. The collection, storage, and use of such data raise significant privacy and security concerns. It is essential to handle geospatial data with utmost care to protect the privacy of individuals and organizations involved.

To address privacy concerns, data anonymization techniques can be employed to remove or alter personally identifiable information. Access controls and encryption methods should be implemented to secure geospatial data during storage and transmission. Furthermore, adherence to legal and ethical guidelines, such as obtaining informed consent and complying with data protection laws, is crucial to ensure responsible handling of geospatial data.

Bias and Fairness

Geospatial data analysis can be subject to various forms of biases, including sampling bias, measurement bias, or algorithmic bias. These biases can lead to unfair or discriminatory outcomes, impacting decision-making processes, resource allocation, and policy implementation.

To mitigate biases, it is essential to critically examine the data sources, sampling methods, and analysis techniques used in geospatial data analysis. Transparency and openness in both data collection and analysis are vital to identify and address any bias present. Additionally, the use of diverse and representative datasets, along with rigorous validation techniques, can help ensure fairness in geospatial data analysis.

Environmental and Social Impact

Geospatial data analysis has the potential to influence environmental and social aspects. Decision-making based on geospatial analysis can impact land use planning, natural resource management, or urban development. It is crucial to consider the potential environmental and social implications of geospatial data analysis.

To address these concerns, sustainability and social justice should be integrated into geospatial analysis methodologies. This involves considering the long-term environmental and social impacts of decisions and incorporating community engagement and participatory approaches. Additionally, adherence to ethical guidelines and frameworks, such as the principles of responsible data use, can help mitigate potential negative consequences.

Open Data and Data Sharing

The availability of open geospatial data has significantly contributed to advancements in geospatial analysis. However, data sharing and openness need to be balanced with privacy, security, and intellectual property rights. There are ethical considerations involved in sharing sensitive or proprietary geospatial data.

To promote open data and data sharing, it is essential to adopt data governance frameworks that ensure responsible and ethical data use. This includes defining ownership and usage rights, establishing data sharing agreements, and protecting intellectual property. Transparency and collaboration among stakeholders can help foster a culture of responsible data sharing and maximize the potential benefits of open geospatial data.

Case Study: Ethical Considerations in Location-based Targeted Advertising

A contemporary example of ethical considerations in geospatial data analysis is location-based targeted advertising. Businesses can use geospatial data to target advertisements based on users' real-time location. While this can enhance the effectiveness of advertising campaigns, it also raises questions about privacy and data consent.

To address these concerns, businesses should obtain informed consent from users before collecting or using their location data for targeted advertising. Transparency and clear communication regarding data collection and usage are essential. Similarly, regulations and guidelines should be established to govern the responsible use of geospatial data in advertising, ensuring that privacy rights are protected.

Resources and Further Reading

Here are some resources and references for further exploration of the challenges and ethical considerations in geospatial data analysis:

- Craglia, M., Haining, R., & Wiles, P. (Eds.). (2014). *Spatial Analysis and GeoComputation: Selected Essays*. CRC Press.
- Goodchild, M. F., & Kemp, K. K. (Eds.). (2012). *Geographic Information Science and Technology Body of Knowledge*. Association of American Geographers.
- Haklay, M., & Weber, P. (Eds.). (2008). *OpenStreetMap: User-Generated Street Maps*. Springer.
- Monmonier, M. S. (2002). *Spying with Maps: Surveillance Technologies and the Future of Privacy*. University of Chicago Press.
- Quinn, G., & Shaw, N. (2013). *Ethics in Public Relations: A Guide to Best Practice*. Kogan Page.
- Sieber, R. E. (Ed.). (2013). *The GIScience Research Agenda*. Springer.

These resources provide detailed insights into the challenges and ethical considerations in geospatial data analysis. They cover a range of topics and perspectives, including data quality, privacy, biases, ethics in advertising, and future research directions.

Resources and Further Reading

In this section, we will provide a list of resources and references for further reading on geospatial data analysis and mapping with Python. These resources will help you delve deeper into the subject, explore advanced techniques, and gain a more comprehensive understanding of the field.

Books

1. **Python for ArcGIS** by Laura Tateosian
   This book provides a comprehensive guide on how to use Python for geospatial analysis within the ArcGIS environment. It covers a wide range of topics, including data management, spatial analysis, and map automation.

2. **Geographic Data Science with PySAL and the PyData Stack** by Serge Rey and Dani Arribas-Bel
   This book explores the use of Python in geographic data science, focusing on the PySAL library. It introduces various spatial analysis techniques, spatial econometrics, spatial regression, and network analysis.

3. **Mastering Geospatial Analysis with Python** by Silas Toms
   This book covers advanced geospatial analysis techniques using Python. It explores topics such as raster analysis, data visualization, network analysis, and geostatistics. It also provides hands-on examples and practical applications.

4. **Python Geospatial Development** by Erik Westra
   This book focuses on using Python for geospatial development. It covers topics like data manipulation, geocoding, spatial indexing, and web mapping. It also

introduces various Python libraries such as GDAL, OGR, and Shapely.

5. **Geoprocessing with Python** by Chris Garrard
   This book provides a comprehensive guide to geoprocessing using Python. It covers topics such as spatial and attribute queries, data transformations, and raster analysis. It also includes practical examples and case studies.

Online Resources

1. **GeoPython** (geopython.net)
   GeoPython is an online resource that provides tutorials, code snippets, and resources for using Python in geospatial analysis. It covers topics such as data acquisition, data preprocessing, spatial analysis, and data visualization.

2. **PySAL Documentation** (pysal.org)
   The PySAL documentation is a valuable resource for understanding and working with the PySAL library. It provides detailed explanations of various spatial analysis techniques, along with code examples and references to academic literature.

3. **GeoPandas Documentation** (geopandas.org)
   The GeoPandas documentation is an excellent resource for working with geospatial data in Python. It provides guidance on data manipulation, spatial operations, and visualization using the GeoPandas library.

4. **QGIS Python API Documentation** (qgis.org)
   The QGIS Python API documentation is a comprehensive resource for utilizing Python in the QGIS software. It provides guidance on automating workflows, creating plugins, and accessing QGIS functionality through Python scripts.

5. **Python Spatial Analysis Library (PySAL) Cookbook** (pysal.org)
   The PySAL Cookbook is a collection of notebooks and examples that demonstrate how to use PySAL for spatial analysis tasks. It covers topics such as exploratory spatial data analysis, spatial regression, and spatial econometrics.

Online Courses

1. **Spatial Data Science with Python**
   This online course, offered on platforms such as Coursera and edX, provides a comprehensive introduction to spatial data science using Python. It covers topics such as geospatial data manipulation, exploratory data analysis, spatial statistics, and machine learning.

2. **Geospatial and Environmental Analysis**
   This online course, offered by the University of California, Santa Barbara on the Coursera platform, focuses on geospatial analysis using Python. It covers topics such as spatial data structures, spatial statistics, spatial data visualization, and geospatial modeling.

3. **Applied Spatial Data Analysis with R**
   Though focused on R rather than Python, this online course offered on platforms such as Coursera and edX provides a comprehensive introduction to spatial data analysis. It covers topics such as spatial autocorrelation, spatial interpolation, point pattern analysis, and spatial regression.

Conferences and Workshops

Attending conferences and workshops is a great way to stay updated on the latest advancements in geospatial data analysis and mapping. Here are some notable conferences and workshops in the field:

- **FOSS4G (Free and Open Source Software for Geospatial)**
  FOSS4G is a global conference focused on open source geospatial software. It features presentations, workshops, and demonstrations on a wide range of topics related to geospatial data analysis, mapping, and visualization.

- **PyCon**
  PyCon is a conference for the Python programming language community. It covers various domains, including geospatial data analysis. Attendees can participate in talks, tutorials, and sprints to enhance their knowledge and skills.

- **GeoPython Conference**
  GeoPython is a conference dedicated to exploring Python libraries and tools for geospatial analysis. It includes workshops, tutorials, and presentations on topics such as GIS, remote sensing, and spatial data processing.

Blogs and Websites

- **GIS.SE** (gis.stackexchange.com)
  GIS Stack Exchange is a question-and-answer website for GIS professionals, analysts, and developers. It is a great resource to find answers to specific geospatial analysis questions and learn from the experiences of others.

- **Towards Data Science** (towardsdatascience.com)
  Towards Data Science (TDS) is an online platform that publishes articles and tutorials on various data science topics, including geospatial analysis. TDS is a valuable

resource for learning about new techniques and approaches in the field.

- **Planet Python** (planetpython.org)
  Planet Python is an aggregator that collects posts from various Python-related blogs. It includes blog posts on geospatial analysis and can be a great source of articles, tutorials, and tips from the Python geospatial community.

- **Real Python** (realpython.com)
  Real Python is a website that publishes practical articles and tutorials on Python programming. It covers various domains, including geospatial analysis, and provides in-depth explanations and hands-on examples.

Online Communities

Engaging with online communities can be an excellent way to connect with individuals who are passionate about geospatial data analysis and mapping with Python. Here are some notable communities:

- **GeoPython Community** (geopython.net)
  The GeoPython Community is an active online community dedicated to geospatial data analysis with Python. It offers a platform for individuals to connect, share knowledge, and collaborate on geospatial projects.

- **GIS Stack Exchange** (gis.stackexchange.com)
  GIS Stack Exchange, mentioned earlier as a valuable resource, also serves as a community where users can ask questions, provide answers, and engage in discussions related to geospatial data analysis.

- **GIS Python Slack Group**
  The GIS Python Slack Group is an online community

where individuals interested in geospatial analysis using Python can communicate, share resources, and seek help. It provides a collaborative environment for learning and problem-solving.

Journals

Journals are an excellent resource for accessing cutting-edge research and advancements in the field of geospatial data analysis. Here are some notable journals:

1. **Transactions in GIS**
   Transactions in GIS is a journal that publishes research articles on various aspects of geospatial data analysis. It covers topics such as spatial statistics, spatial modeling, spatial data mining, and geovisualization.

2. **International Journal of Geographical Information Science**
   The International Journal of Geographical Information Science (IJGIS) publishes original research on all aspects of geospatial information science, including geospatial analysis and mapping. It includes articles on theoretical advancements and practical applications.

3. **Computers, Environment and Urban Systems**
   Computers, Environment and Urban Systems is a journal that focuses on the application of computer science and information systems in urban studies. It covers topics such as spatial analysis, urban modeling, and geospatial data integration.

Additional Resources

- **GitHub** (github.com)
  GitHub hosts numerous repositories related to geospatial

data analysis and mapping with Python. Exploring repositories of open-source projects and libraries can provide valuable insights and code examples.

- **ESRI Forums** (community.esri.com)
  The ESRI Forums are an online platform provided by ESRI, a leading GIS software company. They offer a space for discussions, knowledge sharing, and troubleshooting related to ESRI products and geospatial data analysis tasks.

- **Python Package Index (PyPI)** (pypi.org)
  PyPI is the official repository for Python packages. It offers a vast collection of Python libraries related to geospatial data analysis. Exploring PyPI can help you discover new libraries and tools for your geospatial projects.

Conclusion

This section has provided a comprehensive list of resources and references for further reading on geospatial data analysis and mapping with Python. The suggested books, online resources, online courses, conferences, blogs, journals, and communities will help you enhance your knowledge, explore advanced techniques, and stay up to date with the latest advancements in the field. Remember, learning is a continuous process, and continuously exploring these resources will contribute to your growth as a geospatial data analyst and mapper.

Index

9 781778 903892